Frequency and Time

MONOGRAPHS IN PHYSICAL MEASUREMENT

Series Editor

A. H. COOK F.R.S., F.R.S.E.
Jacksonian Professor of Natural Philosophy,
The University of Cambridge.

P. KARTASCHOFF: Frequency and Time (1978)

Frequency and Time

P. KARTASCHOFF

Telecommunications Department
Research and Development Division, Swiss Post Office,
Bern, Switzerland

1978

ACADEMIC PRESS

London   New York   San Francisco

A Subsidiary of Harcourt Brace Jovanovich, Publishers

ACADEMIC PRESS INC. (LONDON) LTD.
24/28 Oval Road
London NW1

United States edition published by
ACADEMIC PRESS INC.
111 Fifth Avenue
New York, New York 10003

Library of Congress Catalog Card Number: 77-74366
ISBN: 0-12-400150-5

Printed in Great Britain by Page Bros (Norwich) Ltd, Mile Cross Lane, Norwich

Editor's Preface

Measurement is at the heart of life today.

Our knowledge of the physical world depends on measurement, so does our capacity to use that knowledge to change the world by engineering and so does our ability to speak to each other by radio and television. Measurement developed in the early civilizations through the needs of building and trade and it remains a very important function of a government to provide a system of measurement and standards that will enable trade to be carried on both nationally and internationally. However, the precision asked of measurements in trade is commonly much less than that required for developments of our time. Mass production of mechanical and electrical assemblies hangs on the ability to make components with such high precision that they will fit together without further adjustment; it requires machines to work to the necessary precision and measurements to check that they do so. Measurement also enters engineering design. The designer uses the knowledge of the physical world available to him in tables or formulae based on physical measurements and works out designs which he transmits to the workshop or builder in numerical form as drawings or tables or otherwise; if the final product is to work, the initial measurements, the designer's specifications and the mechanic's operations must all be carried out with adequate precision and be referred to the same standards.

It is mechanical measurements that play the greatest role in mass production and it was they that first attained a high precision of one part in a million, in a large part in consequence of the need for accurate measurement in the mass production of armaments during the two world wars. Between the wars the techniques of interferometric measurements in terms of wavelength of light were developed. Measurements of time and frequency can now be made to far greater precision thanks to atomic standards of frequency, and a great deal of modern life depends on them for the synchronization of communication systems, for allocation of frequencies, for navigation of ships and aircraft as examples. Yet again, large scale chemical engineering depends on precise measurements of thermodynamic properties so that processes can be scaled up from the small laboratory or pilot plant to the large industrial complex without some failure. Lastly, our knowledge of the world around us, especially in geophysics and astronomy, has been greatly extended by the means now available for precise measurements of times and distances.

It is only within the last twenty or thirty years that techniques for measurement to a part in a million of quantities other than length have been developed Before that time, almost all techniques of measurement were mechanical and the standards to which they were referred were mechanical. Now much measurement depends on quantum physics and the standards utilize quantum principles. A new subject of *quantum metrology* has grown up. It began with the work of Michelson, Fabry and others on the establishment of wavelengths of light as standards of length, together with the interferometric techniques needed to apply them. With lasers, the scope of interferometric measurements has been vastly enlarged and the reproducibility and convenience of standards greatly improved. The next quantity to be related to quantum phenomena was frequency. All time and frequency services now depend on atomic standards, and radio signals of extremely precisely determined frequency can be received by almost anyone with a fairly simple receiver. Subsequently it came to be realised that other quantities could be related to frequency—for example, length through the speed of light—and a great deal of effort has gone into measuring the relevant physical constants and developing techniques so that, for example, electrical standards may be established in terms of frequency measurements.

The advantages of relating other physical quantities to frequency are considerable. In the first place, the precision is high. More important, the standards become widely accessible. With relatively simple cryogenic apparatus, any laboratory that can receive a standard frequency broadcast can establish its local electrical standard to a higher precision than could be done until recently only by a large standardizing laboratory with a complex system of standard cells and resistances.

Now that the general lines of developments in quantum metrology are clear and notable advances have already occurred, it seems timely to publish a series of books on the methods and standards of precise measurements of today. It is the aim of this series to set forth the principles, show how standards are established and give some idea of the applications of measurements of high precision.

The present book is the first to appear. It is appropriate that it should deal with time and frequency. Measurements of frequency are the key to quantum metrology and atomic standards have been established for some time and techniques of measurement are mature.

A. H. Cook

Cambridge
January 1978

Preface

The purpose of this book on Frequency and Time Measurements is to provide physicists and engineers with a background to the measurement techniques developed during the last twenty years. One of its features is to illustrate the fact that a silent revolution has occurred through the replacement of astronomical time, as known since the dawn of human civilization, by atomic time scales. This is the result of concurrent research and development efforts in quantum physics, which have provided the basic natural phenomena, and electronics which have provided the technical means to open a new era in the science and art of measurement.

Astronomy seems to have lost its dominant position for the definition of time, since the time scales currently in use are all based on the operation of atomic clocks. This apparent loss has however been more than compensated by the availability of improved time scales independent of the Earth's variable speed of rotation. A few years ago, it would not have been conceivable to write a text on timekeeping without at least an introductory treatise on astronomy. In this respect, this book departs from the tradition. This is by no means due to lack of interest but there is no need to try and improve on a chapter of history which is closed, at least for the time being.

When the author was asked to write this book for the new Academic Press series "Monographs in Physical Measurement", no comprehensive and recent textbook on the subject existed and the wealth of new information produced during the rapid evolution, especially in the decade 1960–1970, was widely scattered in numerous journals, conference proceedings and various more or less known reports.

The contents are arranged into eight chapters. As a general introduction, Chapter 1 attempts to show the reader the importance of frequency and time measurements in many fields of science and technology.

The limited available space had mainly to be devoted to the fundamentals and the techniques of precise frequency and time measurement. Therefore, much has been left to the reader's imagination. Nevertheless, a lot of additional information related to applications has been included in Chapter 8.

Chapters 2 to 7 form the main body of the text comprising the basic concepts of frequency stability, stable frequency generation and processing, time scale definitions and generation and various techniques of measurement.

Chapter 8 reviews the wide field of radio signal time and frequency comparison methods, illustrating at the same time some of the most important applications of time–frequency technology.

The author is aware of the impossibility of completely exhausting this subject. Many interesting details had to be left aside for the sake of conciseness and in order to keep this monograph reasonably short. This text owes most to the work of a great number of authors cited in the references. The author wishes to express his sincere appreciation to all his colleagues and hopes that no relevant work has been omitted.

Special thanks are due to David W. Allan of the U.S. National Bureau of Standards for his review of the first draft of Chapter 2, to the C.C.I.R. for the permission to reprint some reports, to Kurt Hilty of the Swiss PTT for his help in compiling Chapter 5, to Mr Willy Klein, Director of the Swiss PTT Research and Development Division, for his continuing interest, to Professor A. H. Cook, editor of this Series, for his many helpful and encouraging suggestions and finally to the author's wife, Sylvia P. Kartaschoff for the preparation of the manuscript.

Bern and Neuchatel, Switzerland   P.K.
January 1978

Contents

List of Letter Symbols

		Introduced and used in Section
r	radial coordinate	1
	$r = T/\tau$, dead time ratio	2.2
R	resistance (ohm)	3.4
$R_y(\tau)$	autocovariance of random variable $y(t)$	2.2
R_1, R_2	clock readings	8.7
s	complex frequency variable in Laplace transform	3.5
$S_y(f)$	spectral density of random variable $y(t)$.	2.2
	Note: spectral densities of other variables are defined in the same way:	
	$S_\phi, S_x, S_{U_0}, S_\omega$ are the spectral densities of ϕ, x, U_0, ω, etc.	2.2, 7.3
$S_{\phi_{12}}$	combined phase noise spectral density for two oscillators, measured on their phase difference	7.3
t	time as independent variable or parameter	2.1
	interaction time	3.3
t_k	starting time of kth sample	2.2
t_s	synchronization delay	4.4
T	period of periodic waveform	1
	time interval between samples taken	2.2
T_r	interaction time	3.3
$T_{1,2}$	time constants	3.5
TAI	international atomic time scale	4.2
$T(t)$	clock reading as a function of reference time	4.4
T_t	clock pulse period	5.3
T_d	display time	5.3
T_s	time of transmission	8.6
U, U_0, etc.	voltage amplitude, d.c. value	2.1
$U(s)$	Laplace transform of $u(t)$	
u	auxiliary variable $u = \pi f \tau$	2.2
$u(t)$	instantaneous signal voltage	2.1
UT	universal time	4.2
UT0,1,2	classes of universal time	4.2
UTC	coordinated universal time	4.2
U	gravitational potential	1
U_T	total relativistic potential in Earth fixed reference frame	1
U_n	peak noise voltage	5.3
U_{WA}	wave analyser input voltage	7.3
V	voltage	1
v	velocity, speed	1

		Introduced and used in Section
$\sigma_y^2(\tau) = \sigma_y^2(2, \tau, \tau)$	two sample Allan-variance	2.2
$\sigma_H^2(N, \tau)$	Hadamard variance	2.4
τ	averaging time, sampling time, duration of observation	2.2
ϕ	longitude	1
	phase angle (degrees or radians)	2.1
	contact potential (V)	3.4
$\phi(s)$	Laplace transform of $\phi(t)$	3.5
ω	angular velocity	1
ω	angular frequency (radian/s)	
$\omega_{p,q}$	angular frequencies associated with the energy levels $W_{p,q}$	3.3

1

Introduction

Measurements in frequency and time are of fundamental importance for all experimental work in science and engineering. This is easily demonstrated by the fact that time as a parameter is present in most equations describing natural phenomena.

In comparison with other measurable quantities such as length, mass or temperature, time can be distinguished by its dynamic nature. Unlike the former, time cannot be held constant, i.e. it runs and cannot be stopped.

A clock can be stopped, it then shows a point or instant on its own time scale, i.e. the time when we stopped it. However, time goes on and if the clock we stopped should happen to be the only one at hand, we would have lost a time scale for ever. If we start the clock anew, it will be running late. By how much? This can only be found out with the help of another clock which was kept going during our simple experiment.

Here we have implicitly used both meanings of the word time, namely time as date (or epoch, see Chapter 4) on a time coordinate axis having a defined origin, and as time interval during which the clock was stopped. In colloquial language, this double meaning of time can be illustrated by the sentences:

(a) "It is time to have a cup of tea"   date
(b) "We had a good time"   interval

even without proceeding to measurements.

Frequency is a quantity closely related to time, sometimes very freely called its "inverse". This is not quite true but the meaning of frequency is derived from the observations of periodic events, i.e. events which repeat at

1

regular intervals (of time), as for instance the oscillation of a pendulum. If constant, this interval is called the period T and the frequency of the oscillation v is indeed the inverse of the period

$$v = \frac{1}{T} \text{(Hz)} \qquad (1.1)$$

measured in hertz (Hz). One hertz is one period or cycle per second. The latter designations were abolished only a few years ago and can still be found in some texts.

A particular outstanding feature of frequency and time is the high accuracy of the basic definition and the precision of the measurements, which during the last twenty-five years have progressed to such a level that they leave all other measurements of physical quantities far behind.

In the International System of units of measurement based on the Convention of the Metre, the fundamental unit is the second, the hertz being a derived unit. Traditionally, the definition of the second was based on astronomical concepts and observation. Until 1956, the second was defined as 1/86 400 of the mean solar day. As the irregularities of the Earth's rotation had become well known since the early 1930s through the generalized use of quartz crystal clocks, as well as improved methods and instruments for astronomical observations, a definition allowing variations with time of the basic unit of measurement appeared no longer tolerable. Thus, in 1956, a new definition was adopted, based on ephemeris time, one second being a fraction of 1/31 556 925·9747 of the tropical year (see Chapter 4). This unit is infinitely stable by definition.

However, the ephemeris second was difficult to determine as several years of observation were required to reduce the probable error to a few parts in 10^9.

In the meantime, decisive new approaches to the problem of defining an invariant unit of time were undertaken in the field of atomic and molecular spectroscopy. Taking advantage of the developments of microwave electronics before and during the second world war (1939–1945), physicists had discovered many atomic and molecular spectral lines in the centimetre wave band. The use of a particular resonance (or quantum transition) as a possibly not variant frequency standard led to the birth of a new device, the "Atomic Clock".[1] The first of these devices was based on an absorption line at 23·87 GHz of the ammonia molecule. Its accuracy was limited to about one part in 10^7 and could therefore not yet compete successfully with traditional means. Yet subsequent experiments with other methods, especially the caesium atomic beam resonator (see Chapter 3) led to rapid progress in the improvement of the accuracy of laboratory frequency standards, namely about 1 order of magnitude every five years during the past twenty years:

1955	10^{-9}
1960	10^{-10}
1965	10^{-11}
1970	10^{-12}
1975	10^{-13}

This rapid evolution led to serious discussions regarding the choice of a new definition of the second, no longer based on celestial mechanics but on quantum mechanics.

In 1967, agreement was obtained and in October of that year, the XIIIth General Conference on Weights and Measures adopted a new definition, which is still valid today and which reads in the official French version:

"La seconde est la durée de 9 192 631 770 périodes de la radiation correspondant à la transition entre les deux niveaux hyperfins de l'état fondamental de l'atome de césium 133."*

It is interesting to note that in this definition the measured quantity is *frequency* and not time. Whilst in the old definition the second was given as a small fraction of a long period, we now have given the second as a large number of very rapid oscillations. As discussed in more detail in Chapter 4, the transition from the old to the new definition has practical consequences on the operation of clocks.

In the old practice, clocks were used to *keep time* (timekeeper, in French: garde-temps), i.e. time was in the stars and the clocks helped to keep it between observations. Hours, minutes and seconds were results obtained by dividing longer periods of time such as the mean solar day or the tropical year.

With the new definition, longer time intervals are built up by successively adding elementary time intervals. Without changing the inner mechanisms of clocks, i.e. the functional systems composed of oscillator–counter–display, clocks have become *time scale generators.*

Obviously, continuity must be preserved between the traditional and the new ways of defining and measuring time. There is a problem with our physical oscillators generating stable periodic oscillations—strict periodicity implies that each successive cycle is an exact copy of each preceding cycle. We can start counting at any time, i.e. there is no defined origin in our time coordinate.

This problem is overcome by means of *synchronization* according to a conventional origin. The latter has been agreed upon, namely on 1 January

---

* An English translation is given in Section 3.3.

1958, 0 h, 0 m, 0 s. The highest possible accuracy, i.e. closest agreement with the defined unit is required only from a few primary laboratory standards. All other users can adjust the frequency of their clock oscillators to the primary standards. Accordingly, the important quality required from these clocks is the *stability* of their frequency. The higher the stability of the oscillator, the better the uniformity of the generated time scale.

Chapter 2 is therefore entirely devoted to the various aspects of frequency stability, especially the characterization of fluctuations by means of statistical measures. Since we deal with very small deviations from the ideal values, normalized quantities are used wherever possible. Symbols and notations are kept in concordance with those used by the IEEE Technical Subcommittee on Frequency Stability and by the US National Bureau of Standards. The statistical measures used in the Fourier frequency domain and in the time domain are spectral densities and the Allan-variance respectively. An oscillator noise model which can be used to describe the fluctuations of practically all presently known oscillators is given in the form of a power series for the spectral density of normalized frequency fluctuations. Whilst the matter presented is mathematically not very difficult, some of the subjects have been controversial at times, as can be seen in some references cited. The field of theoretical modelling is not exhausted yet because there are some difficulties, especially on account of the impossibility of constructing a *realistic* oscillator noise model using only stationary processes. In this context, the problem of clock time prediction should be mentioned, a problem which still is being investigated. Some notes on this subject are given in Chapter 4.

Chapter 3 reviews the techniques of stable frequency generators, including frequency processing techniques as multiplication, division and synthesis. The presentation of the subjects is user-oriented, i.e. it is not intended to help designing and building frequency generating equipment. Such information is abundant and available through the listed references. The purpose is to explain the principles of operation to the user of such equipment in order to facilitate understanding of the strengths and weaknesses of the devices used for his particular applications.

Further details have been given in a few cases when the author has felt some lack of information in the available literature, at the risk of appearing somewhat uneven in the depth of treatment.

Chapter 4 complements the information given in the previous chapter by passing from frequency sources to time scale generation—it is like the step to be taken from the swinging pendulum to the clock, adding gears, hands, dials, weights as well as somebody who rewinds the clock to keep it running.

The importance of international coordination is stressed and general information with reference to the responsible bodies is given. A paragraph on time codes closes the chapter.

Chapters 5 to 7 are mainly devoted to the measurement techniques: the various modes of measurements using counters, phase–time measurements and frequency-domain (spectrum analysis) measurement techniques.

Chapter 8 describes methods of frequency and time comparison via radio signals. This chapter is probably the most closely linked to applications in this text because of the close interaction between *navigation systems* and *time/frequency technology*. Possibilities, advantages and weaknesses of various methods and available means are reviewed and the latest CCIR report 267-3 is reprinted, giving all detailed information available on currently operating radio time signal, navigation and standard frequency services.

As the chapters of this text deal principally with measurements of frequency and time, the relative value of their contents must be weighted against the possible *applications*.

In general physics, the fundamental importance of recent developments in the field of frequency and time is being increasingly recognized in other fields. The recent development of very stable infrared laser sources and the accurate measurement of their frequencies has led to a new situation in length metrology. The frequency of the 3·39 μm CH_4 resonance line can be measured to much higher accuracy than its wavelength in terms of the current definition of the metre. The latter is based on the wavelength of an emission line of krypton 86 in the visible part of the spectrum, namely

$$1 \text{ m} = 1650763\cdot73 \, \lambda^{86}\text{Kr}$$

The wavelength of about 0·605 78 μm is measured in vacuum and the light emission is generated by means of a specially designed gas discharge lamp. The reproducibility of this definition is limited to about ± 1 part in 10^8. The frequency of the 3·39 μm methane resonance can be reproduced within about 1 part in 10^{11} (see Section 3.3).

If the metre definition were based on the methane instead of the krypton line, the corresponding methane wavelength standard could be about 1000 times more accurate than the krypton lamp now in use. Thus, the development of lasers stabilized by saturated absorption, such as that of methane, iodine and other substances still under investigation, raises new questions about the definition of the metre.

Furthermore, the first successful attempts (see Chapter 3, Ref. 31) have been undertaken to measure the frequency of infrared sources such as the methane stabilized laser in terms of the caesium frequency, using the principles of harmonic generation (see 3.4.1) to multiply an accurately known microwave frequency up into the infrared. It is thus possible to measure the frequency v_0 of the infrared source to an accuracy comparable to that of the caesium atomic frequency standard, whereas, for the time being, the measurement of its wavelength λ_0 is limited by the $\pm 1 \times 10^{-8}$ accuracy of the length standard based on the krypton emission line. This in turn limits the accuracy of the determination of the speed of light since $c = \lambda_0 \cdot \nu_0$.

The choice is therefore still open as to whether there will be a new definition of the metre or a definition of the speed of light. In the latter case, the metre would no longer be an independent unit of measurement but a derived one, depending on the second via a defined value for c. As the discussion opened by the recent developments mentioned above will last for some time, the 15th General Conference of Weights and Measures recommended in June 1975 the use of a provisional value for c, namely $c = 299\,792\,458 \text{ ms}^{-1}$.[2]

Other units could be made to depend on the definition of the second, e.g. the Volt. In the a.c. Josephson effect,[3] the frequency ν of oscillations generated in a superconducting weak link is related to the applied voltage V by

$$\nu = \frac{2\,eV}{h} \qquad (1.2)$$

where e and h are the charge of the electron and Planck's constant, respectively. This simple voltage–frequency relation may be assumed to be invariant with time and thus useful as a working standard for the volt, reducing voltage measurements to frequency measurements.

Another fundamental subject in physics involving time and time measurement is relativity. Until recently, the instabilities of clocks and frequency standards used in laboratories were much larger than the corrections required for time measurements carried out with clocks moving about or at rest in locations having different gravitational potential. Therefore, the discussion of relativistic effects was rather an academic exercise than a real requirement for the practising engineer.

With accuracies and stabilities approaching the order of a number of parts in 10^{13}, the consideration of relativistic effects is necessary, especially for applications in space where high relative velocities and gravitational potential values different from those on the surface of the Earth are encountered.

A treatise on relativity being beyond the scope of this book and outside the competence of the author, the reader is referred to existing texts for the fundamentals[4, 5, 6] of relativity and gravitation. Three recent short papers dealing with the special problems of clock synchronization are cited at the end of Chapter 8. One of these, written by Dr Victor Reinhardt of the NASA Goddard Space Flight Center[7] is relatively easy to understand at the engineering level and is the source of the results given below.

Atomic clocks are currently the best available approximations to ideal clocks in the relativistic sense which define "proper time". There is however already a relativistic correction to be made within the device since the electromagnetic field defined with reference to the frame of coordinates of a device (i.e. the laboratory frame) interacts with atoms moving with respect to that frame.

As shown in Section 3.3, first order Doppler shifts are usually almost perfectly cancelled, but the second order Doppler shift remains and has to be considered as a correction term in the evaluation of a standard. If we had the atom at rest, with its natural frequency v_0 as defined, we measure a value v_m by means of the device in which the atoms move with a velocity v_a:

$$v_m = v_0 \left(1 - \frac{v_a^2}{c^2} \right)^{\frac{1}{2}}, \qquad (1.3)$$

c being the speed of light. The correction term v_a^2/c^2 being small compared to unity, we can approximate:

$$\frac{\Delta v}{v_0} = \frac{v_m - v_0}{v_0} = -\frac{v_a^2}{2c^2}. \qquad (1.4)$$

The uncertainty in the determination of the mean square velocity of the atoms interacting with the electromagnetic field is one of the factors limiting the accuracy capability of actual atomic frequency standards.

It can be shown[7] that if this internal correction is made, the clock can be treated as an ideal one, defining a proper time scale within its technical limits of accuracy and stability.

In a set of clocks, spatially distributed within a given frame of reference, the time scale defined by these clocks referred to each other by a set of rules is called *coordinate time*. There is unfortunately a danger of confusion with the term "coordinated time system" which means just a system of synchronized (in time) and adjusted (in frequency) clocks, not necessarily referred to a reference frame (in the relativistic sense). The international atomic time scale TAI (see Chapter 4) is not only a coordinated system of clocks but also a physical realization of coordinate time based on the SI-second as realized at sea level.[8]

The TAI time scale is computed by the Bureau International de l'Heure in Paris (BIH) as a weighted mean time scale currently based on the operation of about 80 caesium clocks located at 18 laboratories and observatories, in Europe and Northern America. The clock times are intercompared daily by means of radio signals (LORAN-C and/or TV, see Chapter 8) and occasionally by clock transport.

Using an Earth-fixed coordinate system as suggested by Reinhardt,[7] two relativistic effects relevant with respect to the current state of the art in clock accuracy and stability are to be considered. One is due to the difference in gravitational potential and often called "gravitational red shift" and the other to the relative movement of a clock with respect to the clocks fixed in the reference frame.

The normalized frequency or rate difference between two clocks located at points having a different potential is given by the relation

$$\frac{\Delta v}{v_0} = \frac{\Delta U_T}{c^2} \tag{1.5}$$

where

$$U_T = U - \tfrac{1}{2}r^2\omega^2 \sin^2 \theta. \tag{1.6}$$

The first term

$$U = -\frac{GM_e}{r}$$

is the gravitational potential and the second term is the centrifugal potential which appears because of the use of the Earth-fixed coordinate system which rotates with the Earth. G is the gravitational constant, M_e the mass of the Earth and r the distance from the Earth's centre of mass. ω is the angular velocity of the Earth's rotation and $\theta = 90° - \beta$ where β is the geographical latitude. This is an approximation sufficient for our purpose. Near sea level, Eq. (1.5) reduces to

$$\frac{\Delta v}{v_0} = \frac{gH}{c^2} \tag{1.7}$$

where g is the total acceleration at sea level (gravitational and centrifugal) and H the height above sea level. Numerically, this corresponds to a normalized frequency change of 1.09×10^{-13} per km of altitude. Thus, clocks located at points of different altitude above sea level run at different rates but this effect is taken into account in the computation of TAI by the BIH.

If two distant clocks fixed at sea level are synchronized using a two way method (see Chapter 8) for determining and correcting for the signal propagation delay, there remains a finite time difference of

$$\Delta T = -\frac{\omega}{c^2} \int_P r^2 \sin^2 \theta \, d\phi \tag{1.8}$$

as an approximation which results from the transform to the coordinate system rotating with the Earth.[7] The integral is computed over the signal propagation path P using the longitude ϕ as variable of integration. In practice, this correction depends only on the locations of the clocks but not on time.

If the two distant clocks are synchronized by means of a transported clock, the coordinate time difference is given by:

$$\Delta T = \int_P \mathrm{d}t \left(\frac{\Delta U_T}{c^2} - \frac{v^2}{2c^2} \right) - \frac{\omega}{c^2} \int_P r^2 \sin^2 \theta \, \mathrm{d}\phi \qquad (1.9)$$

Here the second term is the same as before, whereas the first term includes both effects mentioned above, namely the gravitational redshift ($\Delta U_T/c^2$) and second order Doppler shift $-v^2/2c^2$. The integral is computed over the travel path P using the travel time as integration variable.

In order to obtain an idea about the orders of magnitude involved, let us assume a numerical example as follows. A clock is flown at 10 km altitude above sea level with a constant speed of 1000 km/hour during 6 hours. In the first integral, we then have a gravitational redshift of $1 \cdot 09 \times 10^{-12}$ minus a second order Doppler shift of $4 \cdot 29 \times 10^{-13}$. Altitude and speed being constant, the integral is reduced to a multiplication of these rate corrections by the flight time of 6 h, i.e. $\Delta T = 14 \cdot 3$ ns minus the second integral which is only path dependent and of the order of $\omega r^2/c^2 = 32 \cdot 8$ ns.

The sign of the second integral depends on the east or westward direction of flight. This effect has been tested by means of flying atomic clocks around the Earth in both directions.[9]

At and near the Earth's surface the effects are small and must be taken into account only for very high precision applications. In clocks installed on board of space vehicles, the gravitational shift is much larger. A clock on a satellite in a geostationary orbit will be shifted by about $5 \cdot 4 \times 10^{-10}$ with respect to an Earth-based clock. If the orbit is perfect, no first and second order Doppler shift is observed as the spacecraft is at rest with respect to the Earth-fixed reference frame. For other orbits, the Doppler shifts are to be considered. Furthermore, other coordinate systems may be used, involving appropriate transforms for the definition of synchronization rules.

It is important to note that for using high precision time ordered systems, the theoretical basis is general relativity. This is especially true for distance measurements, e.g. in navigation since we assume the constancy and invariance of the speed of light c (in vacuo) and in the navigation systems now used or proposed, distance measurements are in fact electromagnetic wave propagation time measurements. We thus do not use metre sticks or even tapes and theodolites but signal sources, clocks, counters and the postulate that c is a universal constant. On the other hand, precision clocks could in principle be used for testing some postulates of special and general relativity.

The possibility of testing the gravitational redshift (Eq. 1.5) was brought up already in the very early days of atomic frequency standards over 20 years ago, especially in the context of the rapidly evolving space science and technology. However, up to this time and despite much discussion, studies and laboratory work, no such experiment has been performed. One reason for this delay was the success of the terrestrial experiments made by Pound and Rebka in 1960[10] and Pound and Snider[11] in 1964 confirming the predicted shift to about 1 % using the Mössbauer effect. An experiment using a hydrogen maser oscillator on board a space probe, proposed many years ago by Kleppner, Vessot and Ramsey, has been performed in 1976.[16] Other proposals for carrying a small portable atomic standard along with one of the Apollo flights found no favour with NASA. Also, neither Skylab nor Soyouz carried clocks of sufficient accuracy to perform such an experiment.

Another far reaching proposal made in 1968 by J. Blamont[12, 13] is still somewhere in a waiting line. This experiment consists in launching a deep space probe on a heliocentric orbit. The objective of the mission was to measure the curvature of space in the Sun's gravitational field by means of short laser pulses before and after occultation of the probe by the Sun. An on-board atomic clock and a drag-free system compensating for solar wind and radiation pressure were also included in the projected design of the probe. The drag-free system consisting of a free floating spherical mass located in the probes centre of gravity should allow an orbit determined as perfectly as possible only by gravitational forces. The on-board clock and the laser tracking system should have provided the most accurate method of tracking.

The practical applications of precise time and frequency measurements have been developing in parallel with the improvements in the generation of stable and spectrally pure signals and uniform precise timing systems. The main fields of practical applications can be grouped under the general headings of *navigation* and *communications*.

Navigators have been demanding users of chronometry almost since the times of Christian Huyghens, and the famous story of the prize awarded to John Harrison for his chronometer No. 4 in 1768 is told in many historical books on timekeeping. But it was the development of modern radio-navigation systems which gave great support to the development of more stable oscillators. Determination of position can be based on the measurement of distances and angles or of distances alone. The most accurate positioning systems rely exclusively on distance measurements by means of measurements of the times of arrival of radio waves. Several systems have been developed which are reviewed in Chapter 8 not only as examples for applications but also as means of distant clock time comparison. The systems relying mostly on precise time and frequency are the LORAN-C, OMEGA and the projected NAVSTAR-GPS system.

Around 1967 the interest for time–frequency CAS* was great but there was also a great deal of resistance on behalf of the prospective users due to the eventual high cost of equipment using precision oscillators, especially if it is to be added to other existing equipment still required for navigation, communication and landing aid. A particular problem is the inclusion of small "general aviation" type aircraft in such a scheme since it is inconceivable that the legally required avionics equipment should cost more than the aircraft itself. Time and Frequency CAS has therefore disappeared from the forefront of the scene. Actually, it is very difficult to predict the extent of future use that will be made of time–frequency technology in this field.

What makes precise time and frequency measurements so interesting in navigation is the simple concept of distance measurement by means of a time measurement. Furthermore, the precision requirements are easily stated. Especially for the measurements of distance by means of a time measurement, the need for high precision is almost obvious since a timing error of 1 microsecond produces a distance error of 300 m in a one-way measurement. From this simple rule it is easy to see how important the precision of timing is for the design of any system in which distances are to be measured by means of time measurements.

There is another more complex side to this problem, namely the various ways in which timing errors are translated into the performance of the system. The variety of possibilities to introduce timing into navigation systems and thus the corresponding diversity of requirements are enormous. However, the most stringent requirements are related to defence applications, a field which is not very well suited to public discussion of the really interesting details. On the other hand, there is no limit to individual imagination.

Most of the currently available information can be found in Chapter 8 dealing with time comparison methods using existing navigation systems.

REFERENCES

1. The atomic clock—An atomic standard of frequency and time. *NBS Technical News Bulletin*, **33** (2), 17–24 (1949).
2. 15th General Conference of Weights and Measures, Resolution B (June 1975), recommends use of $c = 299\ 792\ 458\ \text{ms}^{-1}$.

* CAS = Collision Avoidance System.

3. F. K. Harris, H. A. Fowler and P. T. Olsen. Accurate Hamon pair potentiometer for Josephson frequency–voltage measurements. *Metrologia*, **6**, 134–142 (1970).
4. A. Einstein. "Relativity". Crown, New York, 1961.
5. H. Yilmaz. "The Theory of Relativity and the Principles of Modern Physics". Blaisdell, New York, 1965.
6. C. Misner, K. Thorne and J. Wheeler. "Gravitation". Freeman and Co., San Francisco, 1973.
7. V. Reinhardt. Relativistic effects of the rotation of the Earth on remote clock synchronization. *In* "Proc. 6th Ann. PTTI Meeting" (US Naval Research Laboratory, Washington, DC, 3–5 December 1974). NASA Goddard Space Flight Center, Greenbelt, Md, NASA Doc. No. X-814-75-117, pp. 395–424.
8. C.C.I.R. Report 439-1 (Draft revision, 1976).
9. J. Hafele and R. Keating. *Science*, **177**, 166–170 (1972).
10. R. Pound and G. Rebka. *Phys. Rev. Lett.*, **4**, 337 (1960).
11. R. Pound and J. Snider. *Phys. Rev. Lett.*, **13**, 539 (1964).
12. J. Blamont. *In* "Proceedings of the Conference on Experimental Tests of Gravitation Theories", p. 182. JPL Tech. Memorandum 33–499, Pasadena, Calif., 1970.
13. G. M. Israel. *Ibid.* pp. 236–241.
14. H. A. Stover. A time reference distribution concept for a time division communication network. *In* "Proc. 5th Ann. PTTI Planning Meeting". (US Naval Research Laboratory, Washington, DC, 4–6 December 1973). NASA Goddard Space Flight Center, Greenbelt, Md, NASA Doc. No. X-814-74-225, pp. 505–523.
15. J. R. Mensch. Future DCS objectives in communication network timing and synchronization. *Ibid.*, pp. 525–535.
16. To be published in *Metrologia*.

2

Frequency Stability Measures

2.1. INTRODUCTORY REMARKS ON FREQUENCY STABILITY

In this chapter we will discuss some of the forms which are now generally accepted and in which the frequency stability of a generator can be characterized. This discussion is based on the work effected by a large number of contributors during the past decade. The tremendous progress achieved since about 1955 through the introduction of atomic frequency standards, as well as the development of improved quartz crystal oscillators led to some problems in the description of the performance of the new devices. Many applications of stable and spectrally pure signal generators have been made to various fields such as physics, radio-communications, radar, space-vehicle tracking, navigation, etc., but the many definitions and specifications of the required performance are as numerous as the applications. Many experts had difficulties in communicating with colleagues in various adjacent fields of specialization. The most unsatisfactory situation was in the field of so-called "short term stability", where frequency domain and time domain data were investigated and described quite separately. Terms and methods were used which made it difficult, if not impossible, to compare the results obtained by specialists in different fields of application. As an attempt to improve the situation which was at the time rather confused, a special symposium on short term frequency stability was organized in November 1964 by the National Aeronautics and Space Administration and the Institute of Electrical and Electronics Engineers at the NASA Goddard Space Flight Center.[1] After this Symposium, a Subcommittee on Frequency Stability was formed as a part of the Technical Committee on Frequency and Time of the IEEE Professional Group on Instrumentation and Measurement. In May 1966, several members of this subcommittee contributed b original papers to a special issue of the Proceedings of the IEEE.[2] A repor on the characterization of frequency stability was issued in 1970 by th subcommittee mentioned above.[3]

The concepts and methods proposed in that report have been widel accepted by specialists in the field of Time and Frequency. For this reason the following discussion is based mainly on the work of the IEEE Sub committee.

The instantaneous output voltage of a sinusoidal generator is written i the following form

$$u(t) = (U_0 + \varepsilon(t)) \sin(2\pi v_0 t + \phi(t)). \tag{2.1}$$

U_0 is the nominal value of the amplitude and v_0 the nominal value of th frequency.* Both will be assumed as being constant for all further discussion Furthermore, we will assume that the amplitude fluctuation $\varepsilon(t)$ is negligibl in comparison with U_0; $U_0 \gg \varepsilon(t)$.

Whilst it is possible for amplitude fluctuations to be converted into phas fluctuations owing to the action of nonlinear elements, we prefer to leav these special cases aside for the time being. The further discussion is thu greatly simplified. The loss in generality is negligible since such a.m. to p.n conversion can be treated individually in each case.

The instantaneous frequency of the sinusoidal voltage is thus equal to

$$v(t) = v_0 + \frac{1}{2\pi}\frac{d\phi}{dt} = v_0 + v_v \tag{2.2}$$

and is the sum of a constant nominal value v_0 and a variable term

$$v_v(t) = \frac{1}{2\pi}\frac{d\phi}{dt}. \tag{2.3}$$

Throughout this book, we shall use the Greek letter v for the signa frequency. The Latin symbol f will be used as the frequency variable in th representation of spectral densities, i.e. f is used for "Fourier frequency" only This distinction helps avoid any confusion.

Another almost obvious restriction is

$$|v_v(t)| \ll v_0 \tag{2.4}$$

* The angular frequency $\omega_0 = 2\pi v_0$, well-known in electrical circuit theory, is not frequent used here. We prefer the more elaborate form which is often used in books on Fourier Transfor e.g. Blackman and Tukey, "The Measurement of Power Spectra", Dover, New York, 1958.[8]

i.e. we are only interested in reasonably stable oscillators. Large frequency deviation is a subject of frequency modulation theory, and therefore, beyond the scope of this book.

For a more general discussion of oscillators having a wide range of nominal frequencies, normalization is a very useful technique. In our case, the normalized instantaneous frequency offset from the nominal value is designated by $y(t)$ and defined as follows

$$y(t) = \frac{v_v}{v_0} = \frac{1}{2\pi v_0}\frac{d\phi}{dt}.$$ (2.5)

In the literature, the term "fractional frequency offset" for $y(t)$ is frequently used for this normalized quantity. Another useful quantity is the time integral of $y(t)$:

$$x(t) = \int_0^t y(t)\,dt = \frac{\phi(t)}{2\pi v_0}$$ (2.6)

and we have also

$$y(t) = \frac{dx}{dt}.$$ (2.6a)

$x(t)$ is proportional to the instantaneous phase but has the dimension of time. G. Becker[4] has suggested the name "phase-time" (Phasenzeit) for this quantity. In time and frequency technology, it appears logical to measure phase-time expressed in seconds (or microseconds and even nanoseconds) rather than phase expressed in radians or degrees.

The use of the above defined quantities leads to a consistent formalism which is simple to use and has a clear physical meaning since phase-time differences can directly be interpreted as clock-time differences. $y(t)$ and $x(t)$ are to be interpreted as random processes and we shall attempt to describe them by well-known statistical methods. There are excellent textbooks on statistics and random processes;[5, 6, 7] we shall therefore use only as little theory as necessary to describe the relations between some useful descriptions of frequency stability.

2.2. STATISTICAL MEASURES

As stated in the previous section we shall consider the normalized frequency offset from the nominal value, $y(t)$ and its time integral, the phase-time $x(t)$

as random processes. Recalling the definition of $y(t)$ given in Eq. (2.2), one could be tempted to assume that $y(t)$ should have a zero mean over the time of observation, whereas this will certainly not be the case for $x(t)$. However, most real oscillators not only exhibit random frequency variations about a nominal average but also a systematic frequency drift with time, i.e.

$$y(t) = y_r(t) + at + y_0 \qquad (2.7)$$

and also

$$x(t) = x_r(t) + \frac{a}{2}t^2 + y_0 t \qquad (2.8)$$

a being a normalized aging coefficient, y_0 an initial offset and y_r, x_r the truly random processes. In performing a series of measurements over a long enough period of time against a frequency standard, it is always possible to subtract the drift term at and the initial offset y_0 from the data.* We therefore assume that the mean value of $y_r(t)$ over the period of observation is equal to zero.

Existing methods of measurement do not allow the measurement of instantaneous samples of the random process $y(t)$. The result of a frequency measurement is always obtained as an average over a finite time interval τ, any sample $\bar{y}_k$ is of the form:

$$\bar{y}_k(t_k, \tau) = \frac{1}{\tau} \int_{t_k}^{t_k + \tau} y(t) \, dt = \frac{1}{\tau}(x(t_k + \tau) - x(t_k)). \qquad (2.9)$$

There are two main aspects in the analysis of measurement results obtained in the form of such samples $\bar{y}_k(t_k, \tau)$ from the random process $y(t)$; i.e.

Time-domain Analysis
Frequency-domain Analysis

From Time-domain Analysis, quantities such as the *probability distribution* or the *probability density* of the process can be determined. If the probability distribution is *normal*, it is then possible to compute the variance or the standard deviation of a set of samples. If the distribution is found not to be normal, the cause of this behaviour must be investigated. Common causes are, for example, that a systematic drift has not been subtracted or that there

---

* For low-frequency divergent noise spectra, there is a problem in determining y_0, see (Ref. 3). Furthermore, a least square fit to a series $x(t)$ does not result in a correct value for y_0 and is to be avoided (see also Section 4.4).

is some unnoticed perturbation in the measurement setup or in the device being tested.

It is important to note that in such time domain analysis techniques, the independent variable is not the running time t but the sampling time τ. When there is a danger of confusion, it is more proper to speak of the "lag-time domain" when τ is involved as the variable.

Frequency-domain Analysis deals with the problem of knowing the *spectral density* of the process.

It cannot be stated *a priori* which of the two approaches is the better to use. This depends largely on the particular application and also on the limitations of the available measurement techniques. These are discussed in the following chapters where the practical usefulness of each approach and the cases where both can be used are explained and examples are given.

From a theoretical point of view, a fundamental question regarding random processes such as $y(t)$ represented by series of samples $\bar{y}_k(t_k, \tau)$ concerns the stationarity of the process.

Several definitions of stationarity are given in the literature,[6] e.g. wide-sense, strict-sense, autocovariance, stationary increments, etc. The experimenter is confronted with the fact that most of the experimental data he obtains cannot be accurately described by a stationary process. The latter is only a convenient statistical model, born out of the human mind and this model has some properties which are in conflict with what we observe as natural phenomena. The mere fact that any experiment has a beginning and an end is in contradiction with the concept of stationarity.

On the other hand, there are models which are not stationary but useful to describe experimental data, e.g. random walk or flicker noise.

The discussion about the validity of some models is not closed as yet and might continue for some time. The actual problem is perhaps not the question of stationarity of the statistical model, but that of convergence of the integrals used in defining autocovariance and spectral density functions for limited time intervals of observation and limited frequency ranges.

To illustrate this problem, we start from a series of N measurements of duration τ, performed at regular time intervals $T = t_{k+1} - t_k$, with a dead time between measurements of $T - \tau$.

$$\bar{y}_k(t_k, \tau) = \bar{y}_1, \bar{y}_2, \ldots, \bar{y}_N.$$

The first and second moments of the distribution $\bar{y}_k(t_k, \tau)$ are approximated by the mean value

$$\langle \bar{y}_k \rangle_N = \frac{1}{N} \sum_{k=1}^{N} \bar{y}_k \qquad (2.10)$$

and the variance *of the sample* of N values

$$\sigma_y^2(N, T, \tau) = \frac{1}{N-1} \sum_{k=1}^{N} (\bar{y}_k - \langle \bar{y}_k \rangle_N)^2. \tag{2.1}$$

According to the assumption made above, the limit should be

$$\lim_{N \to \infty} \langle \bar{y}_k \rangle_N = 0 \tag{2.1}$$

whereas the limit

$$\lim_{N \to \infty} \sigma_y^2(N, T, \tau) = \sigma^2(\tau) \tag{2.1}$$

should tend to the "true" variance $\sigma^2(\tau)$ of the process if such a limit exis. But this is sometimes not the case. Even after elimination of any initial offs and linear drift, it is found that the sample variance $\sigma_y^2(N, T, \tau)$ depends all of the three variables N, T and τ. The sample variance defined in Eq. (2.1) is therefore not useful to describe experimental data in the time domain.

A possible solution to this problem has been proposed by D. W. All: (Ref. 2, pp. 221–30) and J. A. Barnes[2, 3] who have shown that for limit values of N, T and τ the limit:

$$\langle \sigma_y^2(N, T, \tau) \rangle = \lim_{M \to \infty} \frac{1}{M} \sum_{i=1}^{M} \sigma_{yi}^2(N, T, \tau) \tag{2.1}$$

exists in many cases of interest where the limit of (2.13) does not exist.

The subsequent discussion of the spectral density $S_y(f)$ of $y(t)$ will sho the ways in which the convergence can be obtained. The average of tl sample variance as defined in (2.14) has become known among specialists the time and frequency field as the "Allan-variance". We shall use the broke brackets $\langle \rangle$ also for practical cases where the number of samples is fini but large enough for the error to be small.

The important feature of the definition of the Allan-variance is the restri tion to a limited number N of samples. It is obvious from (2.14) that if tl "true" variance defined in (2.11) diverges for $N \to \infty$, the Allan-variance al diverges for $N \to \infty$.

The behaviour of $y(t)$ in the *frequency domain* is described by its spectr density $S_y(f)$ which is defined in the usual way as the Fourier Transform the auto-covariance $R_y(\tau)$,[8] as follows:

$$R_y(\tau) = \lim_{T \to \infty} \frac{1}{T} \int_0^T y(t') \, y(t' + \tau) \, dt'. \tag{2.1}$$

Then the spectral density is defined according to the well-known Wiener–Khintchine relations as:

$$S_y(f) = 4 \int_0^\infty R_y(\tau) \cos 2\pi f \tau \, d\tau \qquad (2.16)$$

and inversely:

$$R_y(\tau) = \int_0^\infty S_y(f) \cos 2\pi f \tau \, df. \qquad (2.17)$$

The spectral density is a measure of the fluctuation "power" of the instantaneous normalized frequency offset $y(t)$ as a function of the Fourier frequency f. Its dimension is [seconds], as seen from (2.16), since $y(t)$ and $R_y(\tau)$ are dimensionless. If $y(t)$ were a voltage applied to the resistor, then $S_y(f)$ would be an electric power spectral density (see also Ref. 3).

If it is possible to represent $y(t)$ as a fluctuating voltage by some means of measurement, then $S_y(t)$ can be measured by means of a spectrum analyser. The techniques of measurement and their limitations (e.g. finite observation time and resolution bandwidth) will be discussed in Chapter 7.

We note here also some relations between the spectral densities of other quantities defined earlier:

Absolute frequency offset: $\delta v = v(t) - v_0$ [Hz]

$$S_{\delta v}(f) = v_0^2 S_y(f) \quad \text{[Hz]}. \qquad (2.18)$$

Phase:* $\phi(t)$ [radian]

$$S_\phi(f) = \frac{v_0^2}{f^2} S_y(f) \quad [(\text{radian})^2/\text{Hz}]. \qquad (2.19)$$

Phase-time:* $x(t) = \int_0^t y(t) \, dt$ [s]

$$S_x(f) = \frac{1}{(2\pi f)^2} S_y(f) \quad [(\text{radian})^{-2} \text{s}^{-1}]. \qquad (2.20)$$

---

* These relations are not valid for all non-stationary processes, but they hold for the power law spectra discussed in this chapter, as shown, e.g. by A. N. Malakhov.[10]

Angular frequency: $\omega = \dot{\phi} = \dfrac{\mathrm{d}\phi}{\mathrm{d}t}$

$$S_\omega(f) = (2\pi v_0)^2 \, S_y(f) \quad [(\text{radian})^2 \, \text{s}^{-1}]. \tag{2.21}$$

These relations are very often needed for converting measured data as well as to translate formulae between various papers in the very rich literature (see also Section 7.3).

We shall now state an important relation between the spectral density $S_y(f)$ and the general form of the Allan-variance as defined in (2.14):

$$\langle \sigma_y^2(N, T, \tau) \rangle = \frac{N}{N-1} \int_0^\infty S_y(f) \, |H(f)|^2 \, \mathrm{d}f \tag{2.22}$$

where:

$$|H(f)|^2 = \frac{\sin^2 \pi f \tau}{(\pi f \tau)^2} \left(1 - \frac{\sin^2 \pi r f \tau N}{N^2 \sin^2 \pi r f \tau} \right) \tag{2.23}$$

and

$$r = T/\tau. \tag{2.24}$$

This relation has been derived by L. S. Cutler in Annex A of Ref. 3 and is stated here in a form suggested by J. Rutman,[9] who has given a very simple demonstration of the above equation for the special case of $r = 1$; $\tau = T$. We shall not repeat here the detail of the derivations which can be found in the publications cited. It is however worth reminding ourselves that the Wiener–Khintchine relations of (2.16) and (2.17) which are required to derive (2.22) are known to be valid for stationary random processes. Cutler[3] makes two assumptions in his demonstration. The first is that $R_y(t_1 - t_2) = \langle y(t_1) \, y(t_2) \rangle$ exists, namely that $y(t)$ is stationary in the covariance sense and the second assumption is that

$$\langle y^2(t) \rangle = R_y(0) = \int_0^\infty S_y(f) \, \mathrm{d}f \quad \text{exists.}$$

For the second assumption to be true, it is sufficient to assume that $S_y(f)$ is finite in the interval $f_l < f < f_h$ and zero outside this interval, i.e. that there is a lower and a higher cutoff frequency in the device and the measuring equipment.

This condition is always satisfied in practical situations. A problem could remain, namely that the value of the integral could depend in some complicated way upon the values of either cutoff frequency. Further discussion will show that this is not a serious problem.

The postulate of sharp cutoff frequencies f_l, f_h is neither the most general one nor is it very realistic. Any practical measuring equipment always has a high frequency cutoff which is not sharp but may present one of the many known types of low-pass filter characteristics.

The low cutoff-frequency is also always present, not necessarily in the tested device, but in the limited observation time. Fortunately, the integral of (2.22) has the interesting property of existing even for some cases of non-integrable (infinite-power) spectral densities. This is the main reason why the Allan-variance has become a useful parameter for the characterization of frequency stability. (Actually it is a measure of instability, but in the general use of language one tends to avoid negative concepts.)

For the discussion of the convergence properties of the integral of (2.22), it is convenient to substitute variables by setting $\pi f \tau = u$.

Two special cases are interesting:
(a) the limit for $N \to \infty$:

$$\lim_{N \to \infty} \langle \sigma_y^2(N, T, \tau) \rangle = \frac{1}{\pi \tau} \int_0^\infty S_y \left(\frac{u}{\pi \tau} \right) \frac{\sin^2 u}{u^2} \, du \qquad (2.25)$$

(b) the special case of $N = 2$:

$$\langle \sigma_y^2(2, T, \tau) \rangle = \frac{2}{\pi \tau} \int_0^\infty S_y \left(\frac{u}{\pi \tau} \right) \frac{\sin^2 u \sin^2 ru}{u^2} \, du. \qquad (2.26)$$

It is easy to see that the convergence on the lower limit is better for $N = 2$ because of the additional factor of $\sin^2 ru$. The convergence on the lower limit is then obtained if for decreasing f, $S_y(f)$ grows less than $h_{-2} f^{-2}$, h_{-2} being an arbitrary finite constant.

On the higher limit the integral exists if the fastest growing term of $S_y(f)$ is of the form $h_\alpha f^\alpha$ with $\alpha < 1$ even if there were no upper cutoff frequency f_h. On this side the problem is not one of convergence, since in all physically realizable situations there is a high-frequency cutoff. Hence, the problem is the dependence of the result on the actual frequency response of the system.

The "dead-time ratio" $r = T/\tau$ has some particular influence on the value of the integral (2.26). It is easy to show that the average value of the product $\sin^2 u \sin^2 ru$ in the integrand of (2.26) is equal to 1 for $r \neq 1$ but equal to $\frac{3}{2}$ for $r = 1$. The amount of error which is introduced by this singularity again depends on the form of $S_y(f)$ and approaches the maximum of 50 % in the variance computed from measured data only in very special cases. We shall come back to this point in the following section, after the discussion of particular spectra.

The Allan-variance for $N = 2$ has another advantage which has made its application very popular, namely the extreme simplicity of computation from measured data:

$$\langle \sigma_y^2(2, T, \tau) \rangle = \tfrac{1}{2} \langle (\bar{y}_{k+1} - \bar{y}_k)^2 \rangle \qquad (2.27)$$

In a large majority of applications it is possible to make the measurement with negligible dead-time. It then can be assumed that $T = \tau$, i.e. $r = 1$.

The $N = 2$, $\tau = T$ Allan-variance has been proposed as a recommended measure for frequency stability in the time domain. Where there is no risk of confusion, it is designated by

$$\sigma_y^2(\tau) \equiv \langle \sigma_y^2(2, \tau, \tau) \rangle \qquad (2.28)$$

Some authors use the term "Allan-variance" for this special case only. This is not entirely correct, since in the original paper (Ref. 2, p. 221), the general idea of the N-sample variance was developed. A similar expression to that of (2.27) is found in a paper by A. N. Malakhov[10] which deals with the general analysis of signals having non-integrable spectra. He introduces the function $D_x(t_1, t_2) = \langle (x(t_2) - x(t_1))^2 \rangle$ as a special case of a "structure function" of a random process $x(t)$, based on a concept introduced by A. N. Kolmogorov.[11]

W. C. Lindsey and J. L. Lewis[23] have proposed the use of structure functions of the form:

$$D_\phi(\tau) = \langle (\phi(t + \tau) - \phi(t))^2 \rangle$$

and

$$D_{\dot{\phi}}(\tau) = \langle (\dot{\phi}(t + \tau) - \dot{\phi}(t))^2 \rangle$$

as an alternative to $\sigma_y^2(\tau)$ which they criticize as being a biased estimator. The differences between the two sample Allan-variance and structure functions of the above defined form can easily be shown by defining in the same way structure function of the finite time average sample process $\bar{y}_k(t_k, \tau)$ of Eq. (2.9), namely:

$$D_{\bar{y}_k}(\tau) = \langle (\bar{y}_k(t_k + \tau, \tau) - \bar{y}_k(t_k, \tau))^2 \rangle.$$

This is by definition (Eq. (2.7)) equal to:

$$D_{\bar{y}_k}(\tau) = 2\sigma_y^2(\tau).$$

Hence, the two-sample Allan-variance is one half the structure function of the sample process $\bar{y}_k(t_k, \tau)$.

Dealing with the original processes $\phi(t)$, $\dot{\phi}(t)$ or their normalized forms $x(t)$, $y(t)$ is useful and often preferable in theoretical work. However, $y(t)$ is accessible to measurement only via finite sampling times leading to results of the form $\bar{y}_k(t_k, \tau)$. The bias is thus introduced by and inherent to the measurement and has to be taken into account for whatever type of structure function chosen as a time-domain description of measured oscillator behaviour.*

From a limited set of measured data, only a more or less accurate estimate of $\langle \sigma_y^2(\tau) \rangle$ can be calculated, the number M of samples being always limited:

$$\sigma_y^2(\tau, M) = \frac{1}{2(M-1)} \sum_{k=1}^{M-1} (\bar{y}_{k+1} - \bar{y}_k)^2 \tag{2.29}$$

The convergence of this finite sample average towards the theoretical limit $\sigma_y^2(\tau)$ has been investigated by R. C. Tausworthe[12] and P. Lesage and C. Audoin.[13]

From Ref. 12 we learn that for the usually observed forms of spectral density the choice of $N = 2$ is a reasonably good one. Whilst the choice of $N = 3$ or 4 might yield a slightly faster convergence in certain cases, the additional complexity of computation is not very attractive.

The relative standard deviation of the estimated variance:

$$\sigma(\delta) \approx \frac{\sigma[\sigma_y^2(\tau, M)]}{2\sigma_y^2(\tau)} \tag{2.30}$$

with

$$\delta = \frac{\sigma_y(\tau, M) - \sigma_y(\tau)}{\sigma_y(\tau)} \tag{2.31}$$

is shown to improve proportionally to the square root of the number of samples:

$$\sigma(\delta) \approx KM^{-\frac{1}{2}} \tag{2.32}$$

with K being of the order of unity and $M \gtrsim 10$.[13, 24]

Up to this point we have treated the frequency and time domain measures on an equal basis. However, as can be seen in the next section, it is more suitable to formulate an oscillator noise model in the frequency domain rather than in the time domain.

---

* Recent work on structure functions and their relations to Allan-variance and spectral density has been published in References [25] and [26].

Eq. (2.22) shows that it is always possible to compute an Allan-variance from $S_y(f)$ if it is known and if the integral exists. The inverse is possible only within some restrictions and in some cases there is no unique result (see table 2.2). Therefore, *spectral densities such as $S_y(f)$, $S_\phi(f)$, etc., are more fundamental measures* for the characterization of frequency stability than time domain measures such as the Allan-variance.

The reason for which the latter has become very popular is that it is so simple to compute.

Its limits as a spectrum estimator as well as those of some other time-domain data processing methods are reviewed in Section 2.4.

2.3. OSCILLATOR NOISE MODEL

The types of noise observed on the output signal of an oscillator can be represented most suitably by means of the spectral density $S_y(f)$ defined in the preceding paragraph. A simple power-law model of the form

$$S_y(f) = h_{-2}f^{-2} + h_{-1}f^{-1} + h_0 f^0 + h_1 f^1 + h_2 f^2 = \sum_{\alpha=-2}^{2} h_\alpha f^\alpha$$

(2.33)

and

$$S_y(f) = 0 \quad \text{for} \quad f > f_h$$

has been shown by experience to cover all actually known types of oscillators within the limits such as elimination of drift, etc., pointed out before.

The assumption of an ideally sharp upper cutoff frequency f_h in the measuring system seems to be very artificial at the outset. Other types of low-pass characteristics have been used. They lead to more complicated formulae which are more accurate but the corrections involved are often negligible compared to the uncertainties of the experimental results as such. We shall therefore concentrate the discussion on the simpler sharp-cutoff model, introducing the additional corrections only where they are required.

We can then show the relationship between $S_y(f)$ as defined above and $\sigma_y^2(\tau)$ by means of a relatively simple demonstration. For every term of the form $h_\alpha f^\alpha$, we have, using (2.26):

$$\sigma_y^2(\tau) = \frac{2h_\alpha}{(\pi\tau)^{\alpha+1}} \int_0^{\pi\tau f_h} u^{\alpha-2} \sin^4 u \, du$$

(2.34)

that is:

$$\sigma_y^2(\tau) = K_\alpha \tau^\mu \tag{2.35}$$

with

$$\mu = -\alpha - 1 \tag{2.36}$$

and

$$K_\alpha = \frac{2h_\alpha}{\pi^{\alpha+1}} \int_0^{\pi\tau fh} u^{\alpha-2} \sin^4 u \, du. \tag{2.37}$$

For $\alpha < 1$ and $\pi\tau fh \gg 1$, K_α is independent of f_h and τ and becomes just a numerical constant because of the very rapid convergence of the integral. For $\alpha = 1$ and 2, it is obvious that the value of the integral depends on f_h as well as on τ.

A more detailed analysis[3] shows that $\sigma_y^2(\tau)$ can still be written in the form of (2.35), but with $\mu = 2$ and K_α depending on f_h.

The relations for general values of N and r have also been derived in Ref. 3. A slight error in the tables of that reference has been corrected later and the revised tables are given in Refs. 14 and 15. Table 2.1 shows these relations.

The power law expansion of (2.33) is not only convenient but has also some physical meaning. We shall not discuss here the various and numerous causes and sources of noise in actual oscillators but restrict ourselves to a few general remarks. As we shall see in Chapter 3, any oscillator contains a frequency determining element (resonant circuit, quartz crystal resonator, atomic resonator) and a feedback loop comprising an amplifier. Any sources of noise may act either on the frequency or on the phase of the generated signal. It is therefore useful to discuss the noise model in terms of frequency as well as in terms of phase fluctuations. Using (2.20) to transform (2.33) we can define the spectral density of phase-time fluctuations:

$$S_x(f) = \frac{1}{4\pi^2} \sum_{\alpha=-2}^{2} h_\alpha f^{\alpha-2}$$

$$= \frac{1}{4\pi^2} (h_{-2}f^{-4} + h_{-1}f^{-3} + h_0 f^{-2} + h_1 f^{-1} + h_2). \tag{2.38}$$

The various types of noise identified by the α-terms in (2.33) and (2.38)

TABLE 2.1. *Stability Measure Conversion Chart**
(Frequency domain–Time domain)

$S_y(f)$ = one-sided spectral density of y (dimensions are y^2/f), $0 \leqslant f \leqslant f_h$, $f_h = B$, $2\pi f$
$S_y(f \geqslant f_h) = 0$

General definition: $\langle \sigma_y^2(N, T, \tau, f_h) \rangle = \left\langle \dfrac{1}{N-1} \sum_{n=1}^{N} \left(\bar{y}_n - \dfrac{1}{N} \sum_{k=1}^{N} \bar{y}_k \right)^2 \right\rangle$, $\dfrac{dx}{dt} = y = \dfrac{\delta v}{v_0}$,

$r = \dfrac{T}{\tau}$

Special case: $\sigma_y^2(\tau) = \langle \sigma_y^2(N = 2, T = \tau, \tau, f_h) \rangle = \left\langle \dfrac{(\bar{y}_{k+1} - \bar{y}_k)^2}{2} \right\rangle$

Frequency domain (Power law spectral densities)	Time domain (Allan-variances, ...) $\sigma_y^2(\tau)$ $[N = 2, r = 1]$
WHITE x $S_y(f) = h_2 f^2 \left(S_x(f) = \dfrac{h_2}{(2\pi)^2} \right)$ $2\pi f_h \tau \gg 1$	$h_2 \cdot \dfrac{3 f_h}{(2\pi)^2 \tau^2}$
FLICKER x $S_y(f) = h_1 f \left(S_x(f) = \dfrac{h_1}{(2\pi)^2 f} \right)$ $2\pi f_h \tau \gg 1,\ 2\pi f_h T \gg 1$	$h_1 \cdot \dfrac{1}{\tau^2 (2\pi)^2} \left[\tfrac{9}{2} + 3 \ln(2\pi f_h \tau) - \ln 2 \right]$
WHITE y (RANDOM WALK x) $S_y(f) = h_0 \left(S_x(f) = \dfrac{h_0}{(2\pi)^2 f^2} \right)$	$h_0 \cdot \tfrac{1}{2} \tau^{-1}$
FLICKER y $S_y(f) = \dfrac{h_{-1}}{f} \left(S_x(f) = \dfrac{h_{-1}}{(2\pi)^2 f^3} \right)$	$h_{-1} \cdot 2 \ln 2$
RANDOM WALK y $S_y(f) = \dfrac{h_{-2}}{f^2} \left(S_x(f) = \dfrac{h_{-2}}{(2\pi)^2 f^4} \right)$	$h_{-2} \cdot \dfrac{(2\pi)^2 \tau}{6}$

* Adapted from J. A. Barnes *et al.*, "Characterization of Frequency Stability", NE nical Note 394 (October 1970); also published in *IEEE Trans. Instrum. Meas.* IN 105–120 (1971).

Useful relationships:
$$(2\pi)^2 = 39\cdot48$$
$$\ln 2 = 0\cdot693$$
$$2\ln 2 = 1\cdot386$$
$$\ln 10 = 2\cdot303$$

$\langle \sigma_y^2(N, T = \tau, \tau, f_h) \rangle$ $[r = 1]$	$\langle \sigma_y^2(N, T, \tau, f_h) \rangle$
$h_2 \cdot \dfrac{N+1}{N(2\pi)^2} \cdot \dfrac{2f_h}{\tau^2}$	$h_2 \cdot \dfrac{N + \delta_k(r-1)}{N(2\pi)^2} \cdot \dfrac{2f_h}{\tau^2}$ $\delta_k(r = 1) = \begin{cases} 1 \text{ if } r = 1, \\ 0 \text{ otherwise} \end{cases}$
$\dfrac{2(N+1)}{N\tau^2(2\pi)^2}$ $\left[\dfrac{3}{2} + \ln(2\pi f_h \tau) - \dfrac{\ln N}{N^2 - 1}\right]$	$h_1 \cdot \dfrac{2}{(2\pi\tau)^2}\left\{\dfrac{3}{2} + \ln(2\pi f_h \tau)\right.$ $\left. +\dfrac{1}{N(N-1)}\sum_{n=1}^{N-1}(N-n)\cdot\ln\left[\dfrac{n^2 r^2}{n^2 r^2 - 1}\right]\right\}$, for $r \gg 1$
$h_0 \cdot \tfrac{1}{2}\tau^{-1}$	$h_0 \cdot \tfrac{1}{2}\tau^{-1}$, for $r \geq 1$ $h_0 \cdot \tfrac{1}{6}r(N+1)\tau^{-1}$, for $Nr \leq 1$
$h_{-1} \cdot \dfrac{N\ln N}{N-1}$	$h_{-1} \cdot \dfrac{1}{N(N-1)}\sum_{n=1}^{N-1}(N-n)\left[-2(nr)^2\ln(nr)\right.$ $\left. + (nr+1)^2\ln(nr+1) + (nr-1)^2\ln|nr-1|\right]]$
$h_{-2} \cdot \dfrac{(2\pi)^2\tau}{12} \cdot N$	$h_{-2} \cdot \dfrac{(2\pi)^2\tau}{12}[r(N+1) - 1], r \geq 1$

John H. Shoaf, 273.04, National Bureau of Standards, February 1973 usually dominate over some ranges of f and in most cases, not all five terms are significant. The individual terms have been identified by common names given in Table 2.2.

TABLE 2.2

Random walk y:	$\alpha = -2$;	$\mu = 1$
Flicker y:	$\alpha = -1$;	$\mu = 0$
White y		
Random walk x	: $\alpha = 0$;	$\mu = -1$
Flicker x:	$\alpha = 1$;	$\mu = -2$
White x:	$\alpha = 2$;	$\mu = -2$

The most common types of noise observed in oscillators are:[3, 15]

(1) Additive noise: Thermal noise in amplifiers which is simply added to the signal. It manifests itself as phase noise, usually white but band limited at some high cutoff frequency.
(2) Perturbing noise: Thermal and shot noise acting within the feedback loop shows up as white frequency noise[16] and thus as random walk on the phase.
(3) Modulating noise: Random variations of reactive parameters, e.g. semiconductor junction capacities, frequency determining parameters of resonators, etc. These variations are either inherent in the device or due to environmental effects. Flicker x and y usually belong to this class.

The causes of the noise types (1) and (2) are well understood.[1, 2, 3, 16] The causes of flicker noise remain a subject of discussion as long as there is no generally accepted physical model for the mechanism leading to this type of noise. It has been shown[17] that a f^{-1} noise spectrum can be generated by integration of the order of $\frac{1}{2}$ of a white noise process and corresponding computer algorithms have been formulated.[15] It has also been suggested[18] that flicker noise might not be a basic phenomenon in itself but an appearance due to the experimental procedure and data processing methods. However leaving out the $h_{-1}f^{-1}$ term in the model of (2.33) creates serious problems in trying to fit the remaining formula to experimental data. Hence, this approach does not solve the problem. Until more is known regarding the debated existence or non-existence of flicker noise, it appears more useful to bear in mind that the model of (2.33) is only a power-series expansion used to *approximate* real spectral densities and the names have been given to the terms of that power series for sake of convenience. The validity of any model remains subject to the test of application in experimental practice. One is usually satisfied if the use of a model leads to results which are reproducible within the known limits of experimental conditions and if no serious contradictions arise between various experiments.

Errors in the estimation of the spectral density can exceed any limit if the observation time is not long in comparison to the period of low frequency components. Flicker noise and random walk have the undesirable property that the spectral density grows indefinitely for lower and lower Fourier frequencies. The experimental determinations of the corresponding constants is therefore a task which must be approached with caution, whatever method of analysis is applied. There is no mathematical substitute for the lack of data. Experimental investigation of very slow fluctuations therefore remains a long and tedious task.

2.4. SPECTRUM ESTIMATION FROM TIME-DOMAIN DATA

This section deals with the estimation of the spectral density $S_y(f)$ from experimental results given in the form of a time series of sample averages $\bar{y}_k(t_k, \tau)$.

Such results are approximations to samples of the instantaneous normalized frequency offset $y(t)$ only for small values of τ, sampled at a rate $1/T$.

In principle, two methods for the estimation of $S_y(f)$ can be used, namely:

(a) *Discrete Fourier Transform* of a time series of data obtained with small values of τ.

(b) *Analysis of variances* by varying the sampling time, using the known relations between $S_y(f)$ and $\sigma_y^2(\tau)$.

Each approach has its characteristic advantages and drawbacks. Careful consideration of the experimental conditions and the properties of the instrument used is very important if misleading results and false interpretations are to be avoided. The methods mentioned in this section are to be applied mainly for slow fluctuation phenomena (low Fourier frequencies, i.e. lower than a few hertz), where the more straightforward measurements by means of analogue spectrum or wave analysis are not feasible (see Chapter 7).

2.4.1. Fourier Transform

Discrete Fourier Transform (DFT) methods are more general in principle since no *a priori* knowledge has to be assumed about the spectrum to be investigated. However, the methods of variance analysis have found a much wider application than DFT in frequency measurements for the simple reasons that counters are very popular and are required anyway for frequency measurements and that variances and standard deviations are easy to compute.

The fast growing use of automatic data logging equipment and mini computers and especially the availability of Fast Fourier Transform (FFT) algorithms will lead to an increased use of DFT methods.

Literature on this subject is abundant. Introductory review papers have been published by G. D. Bergland[19] on the Fast Fourier Transform and by Paul I. Richards[20] on the computing of reliable power spectra.

A more detailed background is given in the well-known textbook by R. B. Blackman and J. W. Tukey[8] on the measurement of power spectra. We therefore shall not go into much detail here but limit the discussion on discrete Fourier Transform methods to a few general remarks.

Sample averages

The sample averages $\bar{y}_k(t_k, \tau)$ are not instantaneous samples! Even if other precautions mentioned below are taken, the computed spectral density is not $S_y(f)$ but $S_y(f)[\sin^2 \pi f\tau/(\pi f\tau)^2]$. This factor, due to the averaging over the sampling time τ approaches unity only for $\pi f\tau \ll 1$.

Aliasing

The sampling of $y(t)$ at regular intervals separated by the sampling period T produces a sequence of pulses. The spectrum of such a sequence is periodic on the frequency axis with a period $1/T$, i.e. the spectrum is reproduced at every harmonic of the sampling frequency $1/T$. If $y(t)$ contains a finite spectral density above the frequency $f = 1/2T$ (known as the "Nyquist frequency"), the computed spectrum is altered by these high-frequency components. The sampling frequency $1/T$ must therefore be higher than twice the highest frequency f at which the spectrum contains significant power density. If T is limited by experimental conditions, the signal representing $y(t)$ must be filtered before sampling so as to suppress components at frequencies higher than the Nyquist frequency $1/2T$.

Finite data record

The finite amount of data recorded in an experiment of total duration T limits the resolution bandwidth Δf by which possible periodic fluctuations in $y(t)$ could be identified. The finite amount of data also puts a lower limit on the normalized mean square error ε^2 in the computation of $S_y(f)$. According to reference[20] the transform of the entire run of duration T_0 yields the highest resolution $\Delta f \approx 1/T_0$ but a large error in the values of $S_y(f)$. The error ε^2 can be reduced by subdividing the data into M segments of duration

T_0/M and then computing and averaging M sample spectra. However, this is obtained at the expense of resolution Δf which then is approximately equal to $\Delta f \approx M/T_0$. The spectral representation is further complicated by the fact that the Fourier Transform of the finite data blocks of duration T_0/M produces a $[(\sin x)/x]^2$-type spectral window which, besides the main peak has significant sidelobes. This leads to a smearing of the spectrum called "leakage".[19]

Uncertainty relation
Assuming that the mean-square normalized error ε^2 is approximately $1/M$ for an average of M sample spectra, the influence of the finite data length T_0 can be summarized into an uncertainty relation of the form:

$$\varepsilon^2 T_0 \Delta f \simeq 1. \qquad (2.39)$$

From this we may conclude that, for any finite amount of data, the lowest frequency component can be resolved and its amplitude measured with reasonable precision only if many periods of that component are contained in the total duration T_0. This is a very important but often overlooked fact, which also means that a valid analysis of slow fluctuations requires a very long observation time to yield significant results.

The modern trend to get results quickly in order to be the first in publication has a very harmful effect in this field.

2.4.2. Variance analysis

The relation between α and μ of Table 2.2 derived for the power-law spectrum model of oscillator noise suggests that sufficient information about the spectral density $S_y(f)$ can be obtained from time-domain measurements of $\sigma_y(\tau)$ versus τ.

$\sigma_y(\tau)$ is the square root of the Allan-variance. Plotting $\sigma_y(\tau)$ versus τ on double logarithmic paper yields segments of straight lines in the ranges of τ where one of the various terms $h_\alpha f^\alpha$ or $K_\alpha \tau^\mu$ dominates.

This method of describing the time-domain frequency stability of an oscillator has been widely accepted because of its easy application. As with the DFT methods mentioned before, there are some limitations on the use of the Allan-variance as a spectrum estimator which must be considered if erroneous results and interpretations are to be avoided.

First, the relations of Table 2.2 can be used only if the spectral density $S_y(f)$ is of the form given in Eq. (2.33). More specifically, this means that the random fluctuations dominate over any deterministic intentional or spurious phase or frequency modulation of the signal. The most common cause of

such non-random modulation is due to the power line frequency and its
harmonics, i.e. multiples of 50 Hz, 60 Hz or 400 Hz (in aircraft equipment).
The increasingly widespread use of d.c./d.c. converters in battery powered
equipment can also be a source of trouble, in particular the high-efficiency
switching converters which produce signals with high harmonic content.
These possible perturbations can easily be discovered by means of analogue
spectrum analysis (see Chapter 7). Strong interference can be seen on an
oscilloscope.

The second limitation is due to the definition according to Eq. (2.37) and
is also obvious from Table 2.2—both white and flicker phase noise lead to a
$\mu = -2$ dependence. These are dominant for high values of f, hence small
values of τ, and $\sigma_y(\tau)$ shows a slope of -1 on a double logarithmic graph.
It is therefore not possible to distinguish these two types of noise in a $\sigma_y(\tau)$
plot.

Among the attempts made to solve this problem two methods are worth
mentioning:

R. A. Baugh[21] has proposed the computation of Hadamard variance of
the form:

$$\sigma_H^2(N, \tau) = \langle(\bar{y}_{k+1} - \bar{y}_{k+2} + \bar{y}_{k+3} \cdots - \bar{y}_{k+2N})^2\rangle \qquad (2.40)$$

where the terms $\bar{y}_{k+i}$ are defined in the usual way

$$\bar{y}_{k+i} = \frac{1}{\tau}\int_{t_k}^{t_{k+i}} y(t)\,\mathrm{d}t.$$

In the frequency domain, the transfer function of a Hadamard variance
has smaller sidelobes relative to the main peak than the Allan-variance.
Further reduction of the sidelobes is obtained by weighting the term of (2.40)
with binomial coefficients, a technique similar to the known Hamming and
Hanning windows described in.[8] These more elaborate computing techniques are actually special kinds of digital filters. Less computing effort than
with DFT techniques is required but the considerations about aliasing,
leakage and uncertainty mentioned before also apply to this method. A
commercial instrument comprising a counter and a programmed desk
calculator has been introduced in 1976.

The other approach is due to J. Rutman[9] and G. Sauvage.[22] These
authors have shown that by analogue filtering of a voltage proportional to
the phase deviation and measuring this "phase-noise" voltage with an
RMS-voltmeter, a modified variance can be defined. Depending on the type
of filter used, these "high-pass" or "bandpass" variances allow to distinguish
between power law spectral terms of $\alpha = 1$ and $\alpha = 2$ respectively. Based on

analogue measurements, this method is nevertheless mentioned in this section because of the fundamental ideas presented by the authors, namely the discussion of the variance operator in the time domain versus its transfer function operator acting on the spectral density (either of phase or frequency) in the frequency domain.[26]

The two approaches mentioned above are complementary and whilst neither of these methods has yet been applied very often, it is felt that the principles exposed in these papers might be helpful to gain more insight into the complex problems of spectral density estimation by means of simple instruments.

The estimation of spectral density by means of the Allan-variance is less problematic for slow random fluctuations where the white ($\alpha = 0$, $\mu = -1$) and flicker ($\alpha = -1$, $\mu = 0$) frequency noise dominate. In high quality crystal oscillators this is usually the case for $\tau \gtrsim 1$ s.

On the other hand, direct spectrum analysis by means of wave analysers, using the techniques described in Chapter 7, is feasible for Fourier frequencies larger than 1 Hz.

We therefore may conclude that, except for some very special cases, spectrum estimation from time domain data is a useful method for the investigation of slow fluctuation phenomena, i.e. the Fourier frequency domain beneath about 1 Hz. Within the limitations discussed above, spectrum estimation by means of the Allan variance appears to be preferable to other more elaborate techniques in most cases, since its computation is simple and its relation to the spectral density not too involved.

2.5. PHASE-NOISE AND SIGNAL SPECTRUM— APPROXIMATE RELATION

In many practical applications one is interested in "phase-noise" or "phase-time fluctuations" rather than in frequency fluctuations as described by $S_y(f)$. Knowledge of phase noise is very important in telecommunications where stable and clean signals serve as carriers for information. In this context, knowledge of the unmodulated carrier signal spectrum is also of great interest. Phase fluctuations known as "phase jitter" or phase-time fluctuations ("timing jitter") are important parameters in the design of time ordered digital communications systems, navigation and position location systems, etc. The long term evolution of phase time is of interest for the definition of time scales (see Chapter 4).

In this section we will restrict the discussion to the frequency domain above some lower cutoff frequency (a few Hz), i.e. to rapid or "short term" fluctuations. The reason for this restriction is that for arbitrarily low values of f, i.e. for long times of observation, the phase $\phi(t)$ and phase-time $x(t)$ become strongly divergent as can be seen from (2.38) for $S_x(f)$ as well as for $S_\phi(f)$ given by

$$S_\phi(f) = v_0^2(h_{-2}f^{-4} + h_{-1}f^{-3} + h_0 f^{-2} + h_1 f^{-1} + h_2) \qquad (2.41)$$

obtained by applying (2.19) to (2.33). In the time domain, this means that $x(t)$ and $\phi(t)$ cannot be described by a random process having a zero mean value. Only for short times after an initial condition $\phi(0) = 0$, $\phi(t)$ will remain small compared to 1 radian. In a finite frequency range $f_l < f < f_h$, the mean square phase deviation is given by:

$$\langle \phi^2 \rangle_{f_l, f_h} = \int_{f_l}^{f_h} S_\phi(f)\, df. \qquad (2.42)$$

This allows to compute the RMS phase deviation $\langle \phi^2 \rangle_{f_l, f_h}^{\frac{1}{2}}$ for a limited Fourier frequency band between f_l and f_h.

If the RMS phase deviation is small compared to one radian

$$\langle \phi^2 \rangle^{\frac{1}{2}} \ll 1 \text{ radian} \qquad (2.43)$$

then it is possible to use an approximate relation to compute the power density of phase-noise modulation (PNM) sidebands. The following definition has been suggested by D. Halford of the NBS:[14, 15]

$$\mathscr{L}(f) = \frac{\text{Power density, one PNM sideband per Hz}}{\text{Total signal power}}. \qquad (2.44)$$

Called Script $\mathscr{L}(f)$, the ratio of the power in one sideband due to phase modulation by noise (referred to 1 Hz bandwidth) to the total signal power (carrier plus two symmetric sidebands) is a normalized power spectral density. The Fourier frequency f appears in this definition quite logically as an offset from the nominal carrier frequency v_0, i.e.

$$f = v - v_0 \quad [\text{Hz}] \qquad (2.45)$$

Hence, according to (2.44) the normalization condition is:

$$\int_0^{+\infty} \mathscr{L}(v)\, dv = \int_{-v_0}^{+\infty} \mathscr{L}(f)\, df = 1 \qquad (2.46)$$

Provided the validity of the small phase deviation condition (2.43) $\mathscr{L}(f)$ is related to $S_\phi(f)$ in good approximation by:

$$S_\phi(f) \approx 2\mathscr{L}(f) \quad [1(\text{radian})^2] \tag{2.47}$$

The limit of the validity of this approximation is $f > f_l$, where f_l has to be large enough to insure that (2.43) holds via (2.42). In plain language this means that (2.47) can be used for low-noise oscillators and not too small offsets from the carrier. Within a few cycles from and including the carrier (2.47) is not valid.

The definition of $\mathscr{L}(f)$ also neglects amplitude-noise modulation sidebands. Within these restrictions which can all be easily verified by experiment (see Chapter 7), $\mathscr{L}(f)$ is a useful parameter to describe the effect of low-level phase noise on the signal spectrum of an oscillator, especially for the purpose of comparison between different types of oscillators.

REFERENCES

1. "Proceedings of the IEEE–NASA Symposium on Short Term Frequency Stability". NASA-SP-80, November 1964.
2. *Proc. IEEE*, **54** (2), (1966).
3. J. A. Barnes, A. R. Chi, L. S. Cutler, D. J. Healey, D. B. Leeson, T. E. McGunigal, J. A. Mullen, W. L. Smith, R. Sydnor, R. F. C. Vessot, G. M. R. Winkler. "Characterization of frequency stability", NBS Technical Note 394. US Govt Printing Office, Washington, D.C., October 1970; also *IEEE Trans. Instrum. Meas.* **IM–20** (2), 105–120 (1971).
4. G. Becker. Über die Begriffe Phase, Phasenzeit und Phasenwinkel bei zeitabhängigen Vorgängen. *PTB-Mitteilungen*, **81**, 348–352 (1971).
5. A. Blanc-Lapierre and R. Fortet. "Théorie des fonctions aléatoires". Masson, Paris, 1953.
6. A. Papoulis. "Probability, Random Variables and Stochastic Processes". McGraw-Hill, New York, 1965.
7. N. Wiener. "Extrapolation, Interpolation and Smoothing of Stationary Time Series". MIT Press, Cambridge, Mass., 1966.
8. R. B. Blackman and J. W. Tukey. "The Measurement of Power Spectra". Dover, New York, 1958.
9. J. Rutman. Characterization of frequency stability: A transfer function approach and its application to measurements via filtering of phase noise, *IEEE Trans. Instrum. Meas.* **IM-23** (1), 40–48 (1974).
10. A. N. Malakhov. Spectral-correlational analysis of signals with non integrable spectra. *Isvestiya VUZ Radiofisika*, **9** (3), 595–607 (1966).
11. A. N. Kolmogorov. *DAN SSSR*, **32**, 19 (1941).
12. R. C. Tausworthe. Convergence of oscillator spectral estimators for counted-frequency measurements. *IEEE Trans. Comm.*, **COM-20**, 214–217 (1972).
13. P. Lesage and C. Audoin. Characterization of frequency stability: Uncertainty

due to the finite number of measurements. *IEEE Trans. Instrum. Meas.*, **IM-22** (2), 157–161 (1973).

14. J. H. Shoaf, D. Halford and A. S. Risley. "Frequency stability specification and measurement: High frequency and microwave signals", NBS Technical Note 632. US Govt. Printing Office, Washington, D.C., (January 1973).

15. B. E. Blair, Ed. "Time and Frequency: Theory and Fundamentals". NBS Monograph 140. US Govt Printing Office, Washington, D.C., May 1974.

16. L. S. Cutler and C. B. Searle. "Some aspects of the theory on measurement of frequency fluctuations in frequency standards. *Proc. IEEE*, **54** (2), 136–154 (1966).

17. J. A. Barnes and D. W. Allan. A statistical model of flicker noise. *Proc. IEEE*, **54** (2), 176–178 (1966).

18. J. De Prins and G. Cornelissen. Power spectrum, frequency stability and flicker noise, *In* Proceedings of the Frequency Standards and Metrology Seminar". Université Laval, Quebec, Canada, 1971.

19. G. D. Bergland. A guided tour of the fast Fourier transform. *IEEE Spectrum*, 41–52 (July 1969).

20. P. I. Richards. "Computing reliable power spectra". *IEEE Spectrum*, 83–90 (January 1967).

21. R. A. Baugh, Frequency modulation analysis with the Hadamard variance. *In* "Proc 25th Ann. Symp. on Frequency Control", pp. 222–225. US Army Electronics Command, Ft Monmouth, NJ, 1971.

22. J. Rutman and G. Sauvage. "Measurement of frequency stability in time and frequency domains via filtering of phase noise", internal note, Adret Electronics Co, F 78190 Trappes, France (1974); *IEEE Trans.* IM-23 (4), 515–518 (1974).

23. W. C. Lindsey and J. L. Lewis. "Modeling, characterization and measurement of oscillator frequency instability". (Final Report to US Naval Research Laboratory, Contract No. N00014-72-C-0107). U.S. Naval Research Laboratory, Washington, DC, June 1974.

24. P. Lesage and C. Audoin. Comments on characterization of frequency stability: Uncertainty due to finite number of measurements. *IEEE Trans. Instrum. Meas.* **IM-24** (1), 86 (March 1975).

25. W. C. Lindsey and C. M. Chie, "Theory of oscillator instability based upon structure functions", *Proc. IEEE*, **64**, 12, 1652–1665 (1976).

26. J. Rutman, "Oscillator specifications: A review of classical and new ideas", *In* Proc 31st Ann. Symp. on Frequency Control, US Army Electronics Command, Ft. Monmouth NJ 1977 (in press).

3

Standard Frequency Generators and Clocks

3.1. INTRODUCTION

The general properties of precision oscillators, frequency multipliers, dividers and synthesizers will be discussed in this chapter.

Improvements of many orders of magnitude have been achieved during the last twenty years in the accuracy and stability of standard frequency generators as well as in the precision of frequency and time interval measurements. The development of atomic frequency standards is the main feature of this progress but the improvement of quartz crystal oscillators has also been very impressive. Both fields are closely connected and the actual performance of atomic frequency standards could not have been attained without the development of better crystal oscillators.

3.2. QUARTZ CRYSTAL OSCILLATORS

Quartz crystal oscillators are nowadays the most commonly used devices as stable frequency generators. A large variety of models exists, which are made for many applications, from the subminiature ultra low power units used in wrist watches to general purpose units used for frequency control in radio and TV receivers, transmitters, various kinds of timers, clocks, etc., up to high quality oscillators enclosed in temperature-controlled ovens. Furthermore, quartz crystal resonators are used as passive devices in a great variety of filters and delay lines, either as discrete elements or more recently in an integrated form as complex electrode patterns deposited on the surface of crystal plates.

In this book, we shall deal only with the general principles of crystal oscillators applied to high precision requirements, in order to keep the discussion within reasonable limits.

3.2.1. Precision quartz crystal resonators

Quartz crystal (rock crystal, single crystal of SiO_2) is an almost ideal material for the construction of mechanical resonators. The piezoelectric effect of this material, an effect which was discovered in the late 19th century, makes it

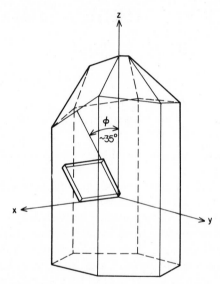

FIG. 3.1. Quartz crystal orientation at AT-cut resonator.

easy to excite mechanical vibrations by means of an oscillating electric field on electrodes deposited on the surface of the crystal. Very low internal friction losses and low temperature coefficients of elasticity and expansion are other important features. Quartz crystals can be obtained either as natural rock crystals or synthetic crystals grown in an aqueous alkaline solution by means of a hydrothermal reaction, both being crystals of high purity with extremely low levels of imperfections.

Fifty years of thorough investigation and refined development have established the optimum ways of designing crystal resonators for various

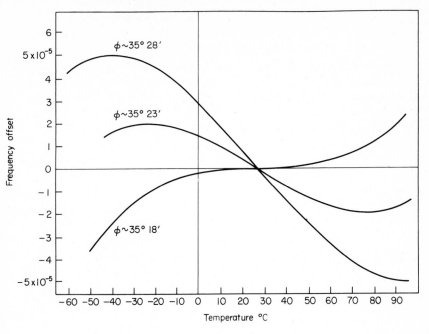

FIG. 3.2. AT-Cut crystals frequency dependence on temperature and orientation angle.

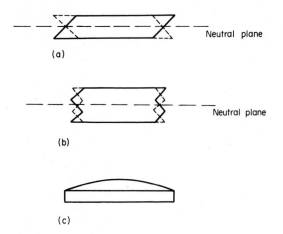

FIG. 3.3. Thickness-shear modes of vibration, motion schematic and exaggerated, (a) fundamental, (b) third overtone, (c) contoured AT-cut, side view.

applications.[1] For high-precision oscillators the optimum design established for some 15 years is the AT-cut lens shaped resonator vibrating in an
overtone (3rd or 5th) thickness-shear mode of vibration. Figure 3.1 shows
the orientation of an AT-cut plate in the natural crystal, in a plane perpendicular to the YZ plane, rotated by 35° away from the Z-axis around the
X-axis. The temperature dependence of the frequency is shown in Fig. 3.2
with the orientation angle as a parameter. X-ray diffraction methods allow
very close control of the orientation in manufacture.

Figure 3.3 shows the mode of thickness-shear vibration:

   (a) the fundamental; and
   (b) the third overtone.

Symmetry allows only the excitation of odd-order overtones. Contour
shaping of the form of a planconvex (Fig. 3.3(c)) or biconvex lens has the purpose of enhancing the excitation of the desired overtone, reducing the influence
of the mount and thereby increasing the Q of the resonator.[2]

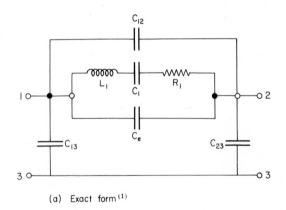

(a) Exact form [1]

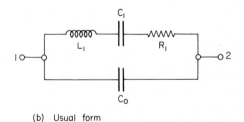

(b) Usual form

FIG. 3.4. Crystal resonator equivalent circuits.

For lowest losses and highest Q, the crystal resonator is mounted inside an evacuated glass or metal envelope. Surface treatment, metal electrode deposition, cleaning and baking (outgassing) methods are all important techniques for achieving high Q and low ageing of the finished resonator.

Near the desired resonant frequency and in the absence of undesired spurious modes of vibration, the electrical characteristics of a quartz crystal resonator as a circuit element can be described by its equivalent circuit shown in Fig. 3.4(a). In this circuit L_1, C_1 and R_1 represent the electric analogue of the mechanical vibrator, L_1 representing the mass, C_1 elasticity and R_1 the friction losses. In the detailed exact circuit C_e represents the capacitance of the electrodes with the quartz dielectric[1, 3] and C_{12}, C_{13}, C_{23} the distributed holder capacities. A simplified version, shown in Fig. 3.4(b) is usually given on specification sheets with C_0 as the two-terminal holder capacitance.

For the design of oscillator circuits, it is important to know all parameters of the circuit (a) since the value C_0 in (b) is not independent of the external circuit parameters.[1]

In precision oscillators, the crystal resonator is operating near its series resonant frequency

$$\omega_0 = (L_1 C_1)^{-\frac{1}{2}}. \tag{3.1}$$

The Q of the series resonant circuit (Fig. 3.4(b)) is approximately equal to:

$$Q = \omega_0 L_1 / R_1 \tag{3.2}$$

if the ratio of C_1/C_0 is so small that the effect of the parallel capacitance C_0 is negligible for the computation of Q. To give an idea regarding the orders of magnitude involved, Table 3.1 shows typical numerical data for 2·5 MHz and 5 MHz fifth overtone crystal resonators. The data for the 2·5 MHz units

TABLE 3.1. Typical data on precision crystal resonators

	2·5 MHz	5·0 MHz
Frequency	2·5 MHz	5·0 MHz
Overtone	5th	5th
Crystal diameter	30 mm	15 mm
R_1	65 Ohms	100 Ohms
L_1	19·5 H	8 H
C_1	$2·1 \times 10^{-4}$ pF	$1·27 \times 10^{-4}$ pF
C_0	4 pF	2 pF
Aging after some months of operation	$<10^{-9}$/month	$<10^{-10}$/day
Operating temperature	50°C	78°C
Quality factor Q	4×10^6	$2·5 \times 10^6$
Max. RMS r.f. crystal current	$<100\,\mu\text{A}$	$<50\,\mu\text{A}$

are taken from the paper[2] by A. W. Warner of the Bell Telephone Laboratories where most of the development of modern high precision crystals took place between 1950 and 1960. The data for the 5 MHz units are typical of commercial units produced by various manufacturers. These are considerably less expensive than the special 2·5 MHz units made for very low ageing, a property which, since the introduction of reliable atomic resonators, has become less important than other characteristics such as low noise, ruggedness and rapid stabilization during warm-up.

The operating temperature must be controlled to much closer tolerances than would appear from the curves in Fig. 3.2. These curves are valid only for thermal equilibrium conditions. Rapid temperature variations of only a few thousands of a degree can cause departures of the frequency of several parts in 10^{10} due to strain effects from nonuniform temperature distribution.

The causes of ageing are mostly known today as being due to mass exchange (absorption and desorption) at the surface of the crystal and its electrodes and to changes in imperfections in the crystal lattice itself. Such defects can also be induced externally, e.g. by nuclear radiation.[1]

3.2.2. High-stability crystal oscillators

A modern high-stability crystal controlled oscillator comprises several elements shown schematically in Fig. 3.5.

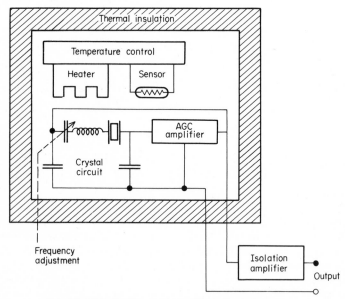

FIG. 3.5. High stability crystal oscillator block diagram.

The crystal resonator is included in the positive feedback loop of an amplifier. The gain of this amplifier is controlled by an automatic gain control system for stabilizing the amplitude of oscillation. The output signal is fed through a buffer amplifier isolating the (variable) load from the oscillator circuit. Modern miniature and integrated circuit technology allow the inclusion of all circuit elements within a temperature-controlled enclosure. Some oscillators are designed with a double concentric temperature control system, having the most critical parts inside an internal oven (crystal resonator and oscillator amplifier) and the remainder of the system in the outer enclosure. Thermal insulation is obtained either with a dewar flask or by means of polymer plastic foam.

Detailed circuit analysis of the crystal oscillator, especially taking into consideration all the possible sources of noise, is complex and beyond the scope of this book. The aspects of noise in crystal oscillators have been treated by W. A. Edson[4] and E. Hafner.[5] More recently the problems of phase flicker noise due to active and passive components have been investigated by D. Halford.[40] Based on this work and on careful testing and selection of components, phase noise density values of

$$S_{\phi} = (10^{-11\cdot8} \text{ radian}^2 \text{ Hz}^2)\frac{1}{f^2} + (10^{-12\cdot6} \text{ radian}^2)\frac{1}{f}$$
$$+ (10^{-14\cdot4} \text{ radian}^2 \text{ Hz}^{-1}). \tag{3.3}$$

have been obtained for 5 MHz precision crystal oscillators designed by H. Brandenberger[33] in 1970. The results constitute an improvement by about 12 dB compared to the previous state of the art. Similar or even slightly better performance has more recently been achieved in some commercial units.

3.3. ATOMIC FREQUENCY STANDARDS

Atomic frequency standards are based on the principles of quantum mechanics which tell us that the interaction of electromagnetic radiation with atoms or molecules leads to interchanges of energy in discrete steps, changing the energy of the absorbing or emitting particle between discrete energy levels. The frequency v_0 of the emitted or absorbed wave packet (photon) is related to the energy difference between levels p and q by the relation

$$hv_0 = W_q - W_p \tag{3.4}$$

where $h = 6\cdot63 \times 10^{-34}$ Js (Js = (Joule × second)) is Planck's constant.

For unperturbed atoms or molecules, the energies W_p and W_q, and, therefore, the frequency v_0 can be regarded as constants, since no experiment has yet shown any secular variation or ageing of these values with time.

Any physical experiment or apparatus devised to measure or generate a frequency v_0 introduces perturbations and errors, not only of a random but also of a systematic nature. The concepts of *accuracy* and *stability* are used to describe the departures of the performance of a practical standard from the assumed ideal.

Frequency stability is discussed in detail in Chapter 2. The concept of accuracy is relevant mainly to the evaluation of "absolute" standards in basic metrology and can be described as follows: the *accuracy* of a primary frequency standard is a measure of its ability to generate a frequency as close as possible to the ideal value v_0.

Obviously, this property can be assessed only by means of a thorough investigation of all known possible perturbing effects. Any potential departure from the assumed nominal value v_0 is called a *bias*. The uncertainty of each bias contributes to the total uncertainty. The results of the investigations of all biases and their uncertainties are usually arranged into an *error budget* and the result of an appropriately weighted combination of the bias uncertainties is an estimate called the *accuracy capability* of the evaluated standard. The task of evaluating the accuracy of a primary standard is at best tedious and time consuming. Furthermore, some duplication of effort in different laboratories is not only desirable but necessary in order to compare the results of independent evaluations. This reduces the danger of unnoticed biases and errors. Ideally the comparisons between laboratories (external estimate) should be in accord with the individual evaluations (internal estimates).

As described above, accuracy is a property which is essential to any basic standard of measurement. However, the recent developments in frequency and time standards have stimulated the elaboration of the concepts of accuracy and stability.

The term "precision" is also frequently used in connection with standards and measurements. Following a quite common practice, we shall use the term "precision" as being a property of *measurements* and *not of standards*. A high precision measurement is one which has low probable error.

Long term stability and accuracy can also be related in a way that the accuracy is always an upper bound for possible long term deviations from the nominal value.

The historical development of atomic frequency standards can be traced as far back as the 1920s[6] when Darwin first linked the quantized orientation of spins in a magnetic field to a resonance phenomenon. In the 1930s the first spectroscopy experiments were made in the microwave range. In 1936

Cleeton and Williams first described the inversion line of ammonia at about 1-cm wavelength and in the same year Rabi at Columbia University developed the atomic and molecular beam resonance techniques for investigations in spectroscopy.

The technological breakthroughs in microwave techniques during World War II led to a very fruitful activity in microwave spectroscopy after the war, and in 1948 Harold Lyons of NBS built the first "atomic clock" based on microwave absorption in ammonia. In the same 5-year period around 1950, the application of atomic-beam spectroscopy to frequency definition and measurement was developed.[6, 7] In 1953 the first maser action was reported, marking the early beginnings of the now wide field of quantum electronics. The first practical application of a caesium beam apparatus was made around 1955[50] and the first commercial caesium beam frequency standards appeared around 1956. In the years from 1955 to 1970, the accuracy capability of caesium beam primary frequency standards has been improved by roughly one order of magnitude every 5 years. Numerous attempts to find better solutions using other techniques and other atoms and molecules were undertaken during the same period. Significant work has been done in Canada, Czechoslovakia, France, Germany, Great Britain, Italy, Japan, the Soviet Union, Switzerland, and the United States.

On the international level, the new definition of the second based on the hyperfine transition of the caesium atom was given final adoption in October 1967 by the General Conference on Weights and Measures, thus legalizing the advent of the era of atomic time. The actual definition[8] of the second reads as follows:

"The second is the duration of 9 192 631 770 periods of the radiation corresponding to the transition between the two hyperfine levels of the ground state of the caesium atom 133".

A standard built and used for the purpose of physically realizing the defined unit is called a primary standard. There are only a few primary frequency standards operated by major national standards laboratories. They all use high performance laboratory designs of caesium beam resonators. Other devices have been tried during the years before and after the adoption of the atomic second definition but none has yet been proved to show a higher accuracy capability. However, one can never exclude the possibility that a better approach could be found, but only the future can tell when this will occur. For such an "X" device to be adopted as a future new definition, independent cross-checks in various laboratories are a necessary condition to any eventual change, as it was the case during the years before 1967.

In this section we will briefly review the principles of operation of those atomic frequency standards which are either in current use or show some real promise for the future. This is done without going into many detailed design features since abundant literature on this subject exists.

Among atomic frequency standards, two main categories can be distinguished, namely the *active* and the *passive* mode of operation.

The active mode is obtained by using the maser or laser principle, i.e. coherent stimulated emission of the radiation within a suitable resonant structure (cavity resonator). In the passive mode, an ensemble of particles (i.e. atoms or molecules) undergoing the desired quantum transition is used as a resonator. An auxiliary source of radiation (commonly called the "slave oscillator") is required to produce the transitions which occur with maximum probability when the frequency of the radiation is, in principle, near v_0. The simplest example is an absorption experiment, but there are various other ways to observe the transitions.

3.3.1. Principles of operation

Three major steps are required in order to implement either the active or the passive mode of operation:

(1) "particle interrogation", which is the way of observing the interactions;

(2) "particle confinement", which keeps the particles in the region of interaction for a sufficiently long time so that the transitions can take place and produce a narrow line;

(3) "particle preparation", in order to obtain a sufficient initial population difference for the observation of a net effect of the transitions. If there were no initial population difference (at least in certain regions), emission and absorption would cancel and no net effect would be observed.

The accomplishment of each of these three steps is associated with several undesired side effects which lead to biases in the generated standard frequency. Fluctuations of these biases and additional sources of noise limit the accuracy as well as the stability of all real devices.

The physics behind these general principles of operation is well documented in the literature.[18] We therefore limit the discussion to a few basic facts summarizing the results of over 20 years of practical experience during which hundreds of proposals were made, experimental investigations conducted and results collected. What we have now is already a result of some natural selection.

The currently most important atomic frequency standards are based on magnetic hyperfine transitions in hydrogen, rubidium and caesium. These transitions are due to spin-spin interactions between the nucleus and the valence electron, in the ground state of the atom. The separation of the energy levels is such that the corresponding frequencies v_0 (Eq. (3.4)) fall into the microwave range, where signals can be generated and detected with conventional and well-known means. One of the characteristics of magnetic hyperfine transitions is the Zeeman effect, i.e. the displacement and splitting of the energy levels in a static magnetic field. The energies of the various levels and their dependence on the applied magnetic field are given by the Breit–Rabi Formula (Ref. 18, p. 80 ff) which, reduced to an approximation sufficient for our purposes, has the following form:

$$W(F, m_F, H) = \frac{-hv_0}{2(2I + 1)} \pm \frac{hv_0}{2}\left(1 + \frac{4m_F}{2(I + 1)}x + x^2\right)^{\frac{1}{2}} \qquad (3.5)$$

where x is proportional to the applied magnetic field H_0:

$$x = \left(\frac{2C_H}{v_0}\right)^{\frac{1}{2}} \cdot H_0. \qquad (3.5a)$$

C_H is the quadratic Zeeman effect coefficient which determines the sensitivity of the standard frequency transition to changes in the magnetic field. I is the nuclear spin quantum number. The magnitude of the total angular momentum of the atom is given by the quantum number F which can assume the values

$$F = I \pm \tfrac{1}{2} \qquad (3.6)$$

the electron spin quantum number being equal to $\tfrac{1}{2}$. The $\pm$ signs in Eqs. (3.5) and (3.6) apply in the same way and describe the two hyperfine levels which are degenerate at zero field. The projection of the angular moment vector F on to the direction of the applied magnetic field vector H_0 is described by the quantum number m_F which according to selection rules can assume the following values

$$m_F = 0, \pm 1, \pm 2, \ldots, \pm F$$

at each F-level. At non-zero field H_0, we have thus a splitting into $2F + 1$ sublevels.

Table 3.2 shows the constants for the hydrogen, rubidium, caesium and thallium atoms required for the numerical evaluation of Eq. (3.5). Figure 3.6 shows the shapes of the energy levels for the values of $I = \tfrac{1}{2}, \tfrac{3}{2}$ and $\tfrac{7}{2}$.

TABLE 3.2. Characteristic constants for ground-state magnetic hyperfine transitions of some currently used atoms

Element	Symbol (isotope)	Ground state	I	v_0 (Hz)	C_H (Hz/Oe^2)
Hydrogen	^{1}H	$^2S_{\frac{1}{2}}$	$\frac{1}{2}$	1 420 405 751	2750
Rubidium	^{87}Rb	$^2S_{\frac{1}{2}}$	$\frac{3}{2}$	6 834 682 608	574
Caesium	^{133}Cs	$^2S_{\frac{1}{2}}$	$\frac{7}{2}$	9 192 631 770	427
Thallium	^{205}Tl	$^2P_{\frac{1}{2}}$	$\frac{1}{2}$	21 310 833 945	20·5

Transitions are allowed between energy levels for which

$$\Delta F = 0, \pm 1$$
$$\Delta m_F = 0, \pm 1. \tag{3.7}$$

It is easy to see from the Breit–Rabi diagrams of Fig. 3.6 that transitions for which $\Delta F = \pm 1$ correspond to microwave frequencies and transitions for which $\Delta F = 0$, $\Delta m_F = \pm 1$ can at low fields be induced by means of much lower frequency signals. In both cases, the probability of spontaneous transitions can be neglected altogether (Ref. 18, p. 121).

Transitions for which $\Delta m_F = 0$ are called σ-transitions and those for which $\Delta m_F = \pm 1$, π-transitions. For σ-transitions induced by an oscillating magnetic field, the transition probability is maximum if the oscillating field direction is parallel to the d.c. field H_0, for π-transitions, if it is perpendicular.

The need for artificially creating a population difference between the levels on which transitions are to be observed is due to the fact, that in a gas in thermal equilibrium at absolute temperature T, the populations N_1, N_2 of two levels with energies $W_1 < W_2$ are related as follows:

$$\frac{N_2}{N_1} = \exp\left[-\frac{W_2 - W_1}{kT}\right] \tag{3.8}$$

where $k = 1·36 \times 10^{-23}$ joules/K is Boltzmann's constant.

For all microwave and lower frequency transitions we have at temperatures used in our devices $hv_{21} = W_2 - W_1 \ll kT$, i.e. the relative population difference is very small. With the magnetic hyperfine transitions shown in Fig. 3.6, it turns out that the effective magnetic dipole moment

$$\mu_{\text{eff}}(H_0) = -\frac{\partial W}{\partial H_0} \tag{3.9}$$

changes its sign for all σ-transitions.

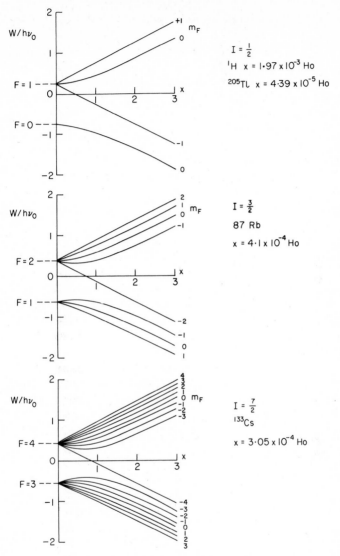

FIG. 3.6. Magnetic hyperfine transitions.

An inhomogeneous magnetic field produces a force F_d acting on the particle:

$$F_d = \mu_{\text{eff}} \cdot \text{grad}\,|H_0|. \tag{3.10}$$

This force is equal to the effective magnetic dipole moment multiplied by the gradient of the magnitude of the magnetic field.

c

Spatial state selection in an atomic or molecular beam is thus possible.

Figure 3.6 shows further that the transitions which are best suited for frequency standard use are the σ-transitions between the two levels for which $\Delta F = \pm 1$ and $m_F = 0$. At low fields H_0, the dependence of the frequency corresponding to this transition is purely quadratic. This follows directly from Eq. (3.5).

$$(H_0) = \nu_0 + C_H H_0^2 \tag{3.11}$$

The values of C_H are given in Table 3.2.

All other transitions show a linear and much stronger dependence on H_0. The coefficients of this linear relation can also be deduced from Table 3.2. In practice these transitions are used for the calibration of the applied magnetic field H_0. A finite d.c. field H_0 must be applied to the region in which the particle interrogation takes place, so that only the desired transitions occur, otherwise accurate measurements would not be possible.

The need for particle confinement can best be illustrated by looking at the result of a computation of the transition probability when the particle is interacting with an electromagnetic field oscillating at or near the frequency corresponding to the energy level separation. The simplest case is a two-level system with energies W_p and W_q, exposed to an oscillating perturbation of frequency $\nu = \omega/(2\pi)$ for a time interval $0 < t < \tau$. This problem has been solved first by I. Rabi and J. Schwinger in 1937; we reproduce here the form given by N. F. Ramsey (Ref. 18, pp. 118 ff) with some abbreviations.

The problem is to solve the time-dependent Schrödinger equation

$$i\hbar \frac{\partial \psi}{\partial t} = (H_0 + V)\psi \tag{3.12}$$

where

$$\psi = C_p(t)\psi_p + C_q(t)\psi_q \tag{3.13}$$

using the Rayleigh–Schrödinger perturbation method. The perturbation is of the form

$$V_{pq} = \hbar b\, e^{i\omega t}, \qquad V_{qp} = \hbar b\, e^{-i\omega t}, \qquad V_{pp} = V_{qq} = 0 \tag{3.14}$$

and (3.12) can be transformed into the following system of linear differential equations

$$i\frac{d}{dt} C_p(t) = \omega_p C_p(t) + b\, e^{i\omega t} C_q(t) \tag{3.15a}$$

$$i\frac{d}{dt}C_q(t) = b\,e^{-i\omega t}\,C_p(t) + \omega_q C_q(t) \tag{3.15b}$$

where $\hbar\omega_p = W_p$ and $\hbar\omega_q = W_q$.

We can assume an initial condition at $t = 0$

$$C_p(0) = 1; \qquad C_q(0) = 0$$

i.e. all atoms considered are in level W_q and none in level W_p. The solution at time t is then:

$$C_p(t) = \left(i\cos\theta\sin\frac{at}{2} + \cos\frac{at}{2}\right)\exp\left[\frac{it}{2}(\omega - (\omega_p + \omega_q))\right]. \tag{3.16a}$$

$$C_q(t) = i\sin\theta\sin\frac{at}{2}\exp\left[\frac{it}{2}(\omega + \omega_p + \omega_q)\right]. \tag{3.16b}$$

The probability of transition from level p to level q is

$$P_{p,q} = |C_q(t)|^2 = \sin^2\theta\sin\frac{at}{2} \tag{3.17}$$

where

$$\cos\theta = \frac{\omega_0 - \omega}{a}, \qquad \sin\theta = -\frac{2b}{a},$$

$$a = ((\omega_0 - \omega)^2 + (2b)^2)^{\frac{1}{2}} \quad \text{and} \quad \omega_0 = \omega_q - \omega_p$$

and by substitution, we obtain:

$$P_{p,q} = \frac{(2b)^2}{(\omega_0 - \omega)^2 + (2b)^2}\sin^2\frac{t}{2}((\omega_0 - \omega)^2 + (2b)^2)^{\frac{1}{2}} \tag{3.18}$$

This result corresponds to the simplest case of interaction between a two level ensemble of atoms to an oscillating field having an amplitude proportional to b during a time t. We see that at resonance, i.e. $\omega = \omega_0$ and $bt = \pi/2$ the transition probability is unity. The linewidth at half amplitude is then approximately equal to $\Delta\nu = 0{\cdot}87/t$, i.e. inversely proportional to the interaction time t.

Without going into a more detailed discussion, which can be found in

Ref. 18, we can also see that the maximum probability as well as the width of the $P_{p,q}$ vs. ω relation depends on the perturbation amplitude. Furthermore, in a real physical system, the interaction time t is not the same for every atom since there is always a velocity distribution determined by gas kinetics and the mechanism of state selection (e.g. in beam resonators).

The purpose of the above example was to show that the transition probability as a function of the perturbation frequency leads indeed to a resonance phenomenon. As with classical resonant systems or circuits, it is possible to define a quality factor $Q = v_0/\Delta v$, i.e. the ratio of the resonant frequency to the half-power linewidth. This Q-factor is very useful to illustrate the behaviour of a standard frequency generator. From the result shown above, the importance of a long interaction time appears clearly. However, it would be dangerous to believe that a high Q-factor is all what is needed to make a good standard. The frequency determining parameters, in our case the energy levels W_p and W_q and the factors perturbing their measurement are equally important.

3.3.2. Particle interrogation

It is only here that the active mode, i.e. laser or maser operation, can be distinguished from the passive mode. In the active mode, a coherent signal output is produced by stimulated emission within a resonant cavity. In the passive mode, the transitions can be observed in various ways:

Absorption;
reemission (amplification of radiation);
detection of particles having made a transition; and
indirect detection (variation of intensity in pumping radiation).

In all these cases, the properties of the slave oscillator play an important role in the total performance of the device. The active maser oscillator frequency stability, for frequencies where $hv \ll kT$, is ultimately limited by the thermal noise in the resonant cavity,[9] with

$$\langle \sigma_y^2 \rangle \approx \frac{1}{Q^2} \frac{kT}{2P\tau}. \tag{3.19}$$

Q is the effective "Q" of the quantum transition and thus directly related to the interaction time which in turn depends on particle confinement. P is the power generated by the device. On the other hand, the frequency stability of a standard operating in the passive mode is limited by shot noise perturbing the frequency of the slave oscillator and can be written as:[12]

$$\langle \sigma_y^2 \rangle \approx \frac{1}{4Q^2} \frac{h}{P\tau} \tag{3.20}$$

where $P = nh\nu$ is the average power exchanged between the radiation field and the particles, assuming that we detect n transitions per unit of time.

It appears, therefore, that for equal Q and equal rate of transitions the passive mode of operation leads to a superior stability, at least in the lower frequency range where $h\nu \ll kT$. Another factor in favour of the passive mode of operation is cavity pulling. In masers and lasers the frequency of oscillation ν is related to the cavity resonance frequency ν_c by:[9]

$$\nu - \nu_0 = \frac{Q_c}{Q}(\nu - \nu_c) = \frac{W}{W_c}(\nu - \nu_c) \tag{3.21}$$

where Q, Q_c, W, W_c are the Q factors and half-power linewidths of the transition and the cavity, respectively. Active oscillators require, furthermore, a high Q_c (low-loss cavity) in order to allow self-oscillation. In passive systems operating far below the level of self-oscillation, the pulling relation[10] can be approximated* by

$$\nu - \nu_0 = \left(\frac{Q_c}{Q}\right)^2 (\nu - \nu_c)\left(2\frac{\Delta P}{P_0}\right) \tag{3.22}$$

where $\Delta P/P_0$ is the fractional departure in excitation power from nominal and there is, at least in principle, no lower limit for Q_c below which the device would fail to operate.

At infrared and optical frequencies, where $h\nu \gg kT$, i.e. in laser oscillators, the frequency is almost entirely determined by the cavity resonant frequency. Lasers can, therefore, practically be ruled out as active frequency standards, but they may well serve as slave oscillators (kind of "optical klystron") locked to a passive resonator.

However, (3.20) sets a limit only for the case of an ideal slave oscillator locked to the resonance. For a complete system and time intervals short compared to the servo response time, the slave oscillator determines the fluctuations of the output frequency.[12] The apparent advantages of a passive system can, therefore, only be realized in practice within the performance limits of available slave oscillators. Passive frequency standards require a servo system controlling the frequency of the slave oscillator to the atomic resonance. Frequency biases due to servo errors are proportional to the

---

* The factor $2(\Delta P/P_0)$ in (3.22) is valid for beam devices such as caesium beams but not for storage devices.[11]

resonance linewidth. With high-Q resonators and careful design of the servo, these errors can be kept smaller than those inherent to the resonator.

3.3.3. Particle confinement

The linewidth (and therefore the Q) of the ensemble of particles (atoms or molecules) interacting with the radiation field is given approximately by the relation

$$W \approx 1/T_r$$

where T_r is the average interaction time (lifetime, relaxation time). The purpose of particle confinement techniques is to make T_r as large as possible, and to reduce uncontrollable motion of the particles with respect to the interrogating radiation field. Three main techniques can be distinguished (in order of increasing sophistication):

> absorption cell;
> particle beam; and
> storage devices: coated storage cell,
> buffer gas cell,
> electromagnetic field containment.

The simple absorption cell is not very suitable for microwave standards, but has regained considerable importance in the infrared and visible region through the introduction of saturated absorption techniques.[13, 14] The very high frequency leads to high-Q values even for short interaction times, together with large signal-to-noise ratios due to the large number of transitions involved.

Particle beam techniques are commonly used in the most accurate microwave frequency standards, i.e. caesium beam resonators. A possible limit of accuracy is the difficulty of knowing the exact velocity distribution which is required to calculate the bias due to second-order Doppler shift.[15] Storage techniques are used in several devices. Coated cells are used for confining hydrogen atoms in hydrogen masers and, more recently, hydrogen storage beam resonators. Buffer gas cells are used in connection with optical pumping[16] in rubidium frequency standards and magnetometers.

Ion storage in electromagnetic fields has been used in high-resolution spectroscopy;[17] however, an interrogation method competitive with other approaches still remains to be developed. There are also problems in determining the mean square velocity of the stored particles and hence the uncertainty of second-order Doppler shift.

The main problems in storage devices using wall coatings or buffer gases are the biases introduced by collisions, which in the case of rubidium gas cells are so large that they preclude their use as primary standards and introduce some ageing. Wall collision shifts in hydrogen storage devices limit the accuracy of those standards.

3.3.4. Particle preparation

As mentioned in 3.3.1, the creation of a sufficient difference in energy level population is necessary only for low-transition frequencies (microwave region) where $hv \ll kT$. At infrared and optical frequencies, the population difference in a gas at thermal equilibrium is sufficiently large that absorption experiments can be done efficiently. The most important methods of particle preparation currently in use are spatial state selection in a beam,[18] optical pumping,[16] and particle collisions.[17, 19]

State selection in a beam can be obtained by means of inhomogeneous magnetic or electric fields interacting with magnetic or electric dipole moments of the particle used. The existence of a dipole moment, however, implies the dependence of the energy on the applied field, i.e. Zeeman and Stark effects, which lead to bias and require careful shielding and control of the applied fields. Magnetic hyperfine transitions in hydrogen and in alkali metal atoms provide energy levels which have a purely quadratic Zeeman effect and negligibly small Stark effect, so that at low magnetic fields the dependence on the field strength is small.

At high fields (> 3000 A/cm), the magnetic dipole moment approaches 1 Bohr magneton (9.27×10^{-24} J/T) and is of opposite sign for the two levels of interest, so that state selection by magnetic beam deflection or focusing of the higher level by multipole magnets becomes easy and relatively efficient.

Optical pumping methods are usually much more efficient than beam state selection provided that an appropriate spectral intensity of the pumping light source can be obtained. In optically pumped alkali vapour cells, microwave transitions can be observed with high signal-to-noise ratios through the change in absorption of the pumping light. However, the simultaneous application of the pumping light and the microwave interrogating field leads to coupling between transitions and thus to biases depending on the spectral characteristics of the pumping light, which are difficult to control.

Population enhancement of certain energy levels by particle collisions is widely used in gas lasers. The same technique is also used in ion storage experiments.

3.3.5. Types of frequency standards

The following types of standard frequency sources have reached some maturity and some show considerable promise for the future:

caesium-beam resonators;
hydrogen masers;
hydrogen storage beam tube;
methane saturated absorption cells; and
rubidium vapour cells.

Furthermore, quartz-crystal oscillators, without being accurate frequency standards, are of high importance as slave oscillators in all radio-frequency standards.

Caesium-beam resonator (Fig. 3.7). The caesium atomic beam is produced from a heated oven containing liquid caesium. The first state selector allows only atoms of a selected hyperfine state to pass through the interaction

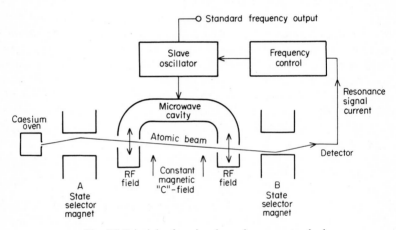

FIG. 3.7. Principle of caesium beam frequency standard.

region, where transitions are produced by a pair of separated oscillating fields (Ramsey method[18]) in a microwave cavity ($\nu_0 = 9\cdot2$ GHz). (Some caesium beams use both emission and absorption, but in this configuration atoms in the two states are spatially separated.) Simultaneously a uniform weak magnetic field (C-field) is applied in order to separate the different sublevels of the hyperfine state, so that transitions occur only between the two levels ($m_F = 0$), where the Zeeman effect is purely quadratic. The second state selector allows only atoms that have completed the transition to the other hyperfine state to be detected. The caesium atoms are detected by surface ionization on a hot-metal ribbon. Usually the ion current is amplified by

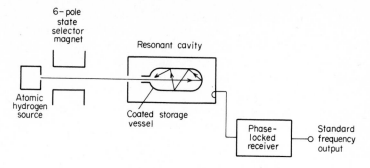

Fig. 3.8. Principle of hydrogen maser frequency standard.

means of a secondary electron multiplier. As a function of excitation frequency, the output current shows a sharp resonance peak.

Hydrogen masers (Fig. 3.8). Atomic hydrogen is produced in an r.f.-discharge source and the beam is formed in a collimator. Atoms in the upper hyperfine state are focused into a coated storage bulb by means of a multipole magnet. The storage bulb is located in the uniform r.f. field region of a high-Q cavity. A weak uniform "C"-field is applied for the same reason as in a caesium tube. The condition of oscillation depends on various relaxation processes, the flux ratio of atoms in the desired and undesired states, and the loaded Q of the tuned cavity.

Hydrogen storage beam tubes (Fig. 3.9). This is a new device[8] based on old ideas,[20] which combines the advantages of the passive resonator with those of hydrogen atom storage in a coated bulb. Source and first state selector are

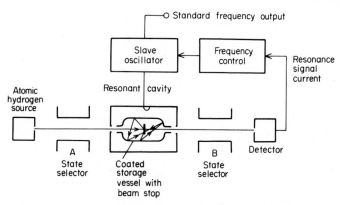

Fig. 3.9. Principle of hydrogen storage beam frequency standard.

similar to those of a hydrogen maser. An output collimator is added to the storage bulb, means for avoiding direct transit of atoms are provided, and a second state selector and hydrogen atom detector are added. The future of this device depends on the development of a high-efficiency detector. In principle, its feasibility has been demonstrated.[21]

Rubidium vapour cell (Fig. 3.10). A light beam from an [87]Rb lamp, filtered by a vapour cell containing [85]Rb, depopulates one of the two hyperfine levels of

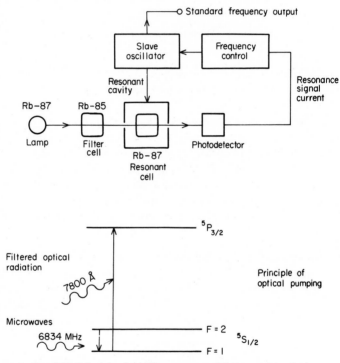

FIG. 3.10. Principle of rubidium vapour cell frequency standard.

[87]Rb atoms stored in another cell which is located inside a cavity. As shown on the simplified energy level diagram, the resonance radiation of the [87]Rb lamp is filtered by means of the [85]Rb isotope absorption in such a way that there are more transitions from the $5S_{\frac{1}{2}}$ $F = 1$ level up to the $5P_{\frac{3}{2}}$ level than from the $5S_{\frac{1}{2}}$ $F = 2$ level. The lower groundstate hyperfine level $5S_{\frac{1}{2}}$ $F = 1$ is thus depopulated. Absorption on the 7800 Å wavelength then ceases until the application of the 6834 MHz microwave radiation to the cavity.

The latter repopulates the lower level by means of induced transitions from the $F = 2$ down to the $F = 1$ level and absorption of the 7800 Å pumping radiation is again observed at the photodetector, which thus provides the resonance signal for locking the slave oscillator. With high levels of pumping light, appropriate choice of buffer gases,[22] and a high-Q cavity, maser oscillation can be obtained. In practical applications, however, the passive mode of operation is the only one used. This is the simplest and least expensive type of atomic frequency standard and it has found wide acceptance for those applications where its ageing with time is no inconvenience. An even simpler variety, using natural rubidium in a resonance cell without additional filter cell, has recently been described.[23]

Methane saturated absorption cell (Fig. 3.11). This very recent development[13] is included in this review because of its potential future as a high-accuracy

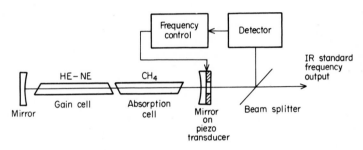

FIG. 3.11. Principle of Methane saturated absorption frequency standard.

frequency standard in the infrared region and as an example for the potential applications of frequency metrology in the definitions of other physical quantities.

A gas cell filled with methane is mounted with a ^{3}He–^{20}Ne gain cell between the two mirrors of a laser cavity. The He–Ne laser can be made to oscillate at a frequency of approximately 88 THz (wavelength 3·39 μm) which coincides with a resonance of the methane molecule. With appropriate methane pressure, strong absorption occurs over the whole range of oscillation of the laser. Methane molecules in the cell interact with both running waves forming the standing wave pattern in the Fabry–Perot resonator of the laser system.

Molecules having arbitrary velocities are perturbed at two different frequencies because of the first order Doppler shift which is of opposite sign for each running wave.

For the molecules moving in a direction perpendicular to the laser beam, the Doppler shift vanishes. These molecules are perturbed by a signal at twice the amplitude without Doppler shift. If the intensity of the laser radiation is sufficiently high and its frequency adjusted, the lower energy level of the molecular transition is depopulated. Additionally, photons will be re-emitted coherently through induced transitions. Therefore, less energy will be absorbed from the laser beam if its frequency is that corresponding to the molecular transition. This phenomenon is called saturation and the dip in the absorption line profile known as the "Lamb dip".[24]

The transition probability depends strongly on the amplitude and frequency of the perturbation as can be seen from Eq. (3.18) (at least in principle because there, we have only one perturbing field). Therefore, the linewidth is very narrow and determined principally by the interaction time. At low gas pressures, the mean free path is longer than the beam diameter, therefore, the interaction time is the flight time through the beam diameter. At room temperature, linewidths of the order of 100 kHz can be observed. At the methane frequency of 88 THz, this corresponds to a Q-factor of roughly 10^9.

The resonance signal is observed by means of a photodetector. The frequency of the He–Ne laser is modulated for scanning the resonance by means of a piezoelectric transducer moving one of the laser mirrors.

The average frequency is locked to the methane resonance by means of an electronic servo system.

3.3.6. Performance

In comparing the performance between the various types of atomic frequency standards a distinction must be made between the highly sophisticated laboratory devices used in national standards laboratories and commercial instruments. The former are operated by experts, evaluated at regular intervals and still subject to improvements, but they are not directly accessible to the great majority of users except through standard frequency and time dissemination services.

The commercial instruments are available to anybody who can afford one. In recent years caesium beam and rubidium cell standards have become available at prices which are no longer considered as prohibitive by many potential users. Rough estimates indicate that from 1965 to 1974 at least several hundred if not 1000 caesium beam standards have been sold (average unit price: about $15 000) and most of them are still operating throughout the world.

The widespread availability of high-performance frequency standards has led to increased demands on the methods of time and frequency dissemination (see Chapter 8) used to relate a local standard to some higher accuracy reference.

In a commercial type frequency standard, accuracy is not the prime design objective, since other factors, such as size, weight, environmental behaviour reliability, power consumption and cost (initial investment as well as operation costs) have an influence on the choice of the type of standard and its application. For the user, this choice is not only a matter of device design and characteristics but rather one of systems design and optimization.

At present, caesium beam devices have an almost unique position in the field of metrology insofar as the principle is used not only in national standards laboratories as the primary reference but also in commercially available instruments. The difference is not in the principle but in the accuracy and the intended use. Thus most caesium standards, whilst having the properties of primary standards within their proper accuracy limits, are operated as secondary standards of time and frequency. In these applications use is made of their long term frequency stability which is superior to that of all other currently available oscillators.

TABLE 3.3. Accuracy capability of present and potential future primary frequency standards

Type	Experimental estimates	Potential limits	Major causes of bias
Caesium beam	2×10^{-13} [27]	5×10^{-14}	Cavity phase shift, "C"-field, Second order Doppler shift. Excitation spectrum[10, 27, 28, 38]
Hydrogen maser	2×10^{-12} [29]	1×10^{-12}	Wall collision shift, Cavity pulling[7, 29]
Hydrogen storage beam	Not yet available	1×10^{-12}	Wall collision shift, Excitation spectrum[8, 29]
Methane cell	1×10^{-11} [13]	1×10^{-12}	Photon recoil,[30] Excitation spectrum, Pressure shift[13]

Table 3.3 shows the 1974 state of the art in the accuracy capability of present and potential future primary frequency standards.

The experimental one-sigma estimates of accuracy capability are all based on published results and the corresponding references are indicated in the table. In the last column the major effects limiting the accuracy and, to some degree, also the long-term stability are mentioned in order to identify the areas of current and future research.

Figure 3.12 shows a time-domain stability plot for caesium-beam, hydrogen maser and methane saturated absorption cell laboratory devices. The established accuracy capability estimates from Table 3.3 are also indicated for comparison. The best stability values for sampling times up to about 10^6 s is obtained with hydrogen masers. For longer periods no data are yet available.

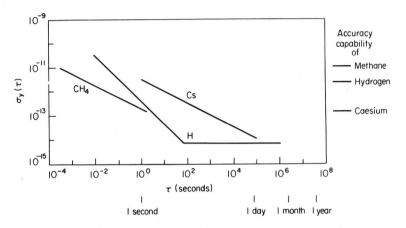

FIG. 3.12. Stability and accuracy of laboratory frequency standards: the data are based on experimental results reported in Ref. 14 for methane, Ref. 26 for hydrogen and Ref. 38 for caesium.

The most probable cause of the absence of published long-term data on hydrogen masers is that commercial availability has been sporadic (Varian Associates Model H-10, 1964–1967; Hewlett-Packard Prototype, 1967). Additionally, hydrogen masers have been built in various laboratories where continuous operation over very long periods was neither possible nor intended.

Early reliability problems have been overcome in well engineered designs such as those built at the NASA Goddard Space Flight Center,[32] the Jet Propulsion Laboratory and the Smithsonian Astrophysical Observatory.

Size, weight and high cost (about five times the price of a commercial caesium standard) have limited the application of hydrogen masers to special cases where its excellent medium-term stability is essential, e.g. for Very Long Base Radio Interferometry[25] and Deep Space Communications.[26]

The methane saturated absorption cell is interesting in two respects, namely as a frequency standard in the infrared region and as a possible candidate for a new definition of either the metre or the speed of light. Its accuracy capability as a frequency standard is still under investigation and it is not known if it will surpass that of microwave devices but as a standard of wave-length it is superior to the Krypton-86 lamp on which the present definition of the metre is based. The successful frequency multiplication from micro-wave frequencies up to the methane frequency[31] has allowed a new deter-mination of the speed of light. The question is still open about how the three quantities, second, metre and the speed of light will be related in the future.*

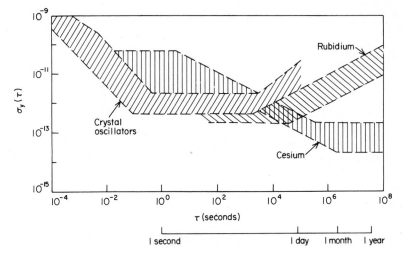

FIG. 3.13. Stability data on commercial frequency standards indicating the margin between specified data and performance of selected units.

The ranges of time-domain stability measures of commercially available frequency standards is shown in Fig. 3.13. The more conservative (upper bound) limit corresponds to manufacturers specifications shown on their data sheets (1972). The best (lower bound) values correspond to measure-ments made on selected units in good laboratory conditions.

For crystal oscillators, the best available values on short-term stability are those reported by Brandenberger, Hadorn, Halford, and Shoaf.[33] The indicated stability data of rubidium vapour frequency standards are those reported by Throne[34] for values of $\tau < 10^4$ s; the longer term data are esti-

* In 1975, the 15th General Conference of Weights and Measures has voted a Resolution to recommend the use of the value $c = 299\,792\,458$ m/s for the speed of propagation of electro-magnetic waves in vacuum.

mated from the expected ageing of the rubidium cell system. The best values for caesium beam standards are based on data reported by Hyatt, Throne, Cutler, Holloway, and Mueller[35] for values of τ up to 10^5 s. The longer term values are based on data reported by Winkler, Hall, and Percival[36] and by Allan.[37]

Standard frequency generators used in clocks and timing systems must not only have the best possible long term frequency stability but also be very reliable since any momentary failure immediately leads to the loss of the time scale.

Reliability can be described with statistical methods and terms such as mean time between failures (MTBF), mean time to repair (MTTR), etc., but it is evident that such data can be established only *a posteriori*. Proof of reliability can therefore be expected for devices which have been in practical use for some time but not for the latest designs. In times of rapidly evolving technology it is difficult to find the best compromise between high performance and reliability. Cost becomes a limiting factor if one attempts to improve the reliability by redundancy, i.e. operating several devices simultaneously. This is more expensive not only because of the increasing number of devices in operation but also because of the growing complexity of the system. In the current state of the art, a MTBF of over 25 000 h can be expected for caesium beam frequency standards. The author has experience of a caesium beam tube lasting for more than seven years. Rubidium frequency standards seem to have similar, though less well documented reliability. Crystal oscillators seems to last much longer, some over twenty years of continuous operation.

If we assume similar electronics technology and design practice, MTBF is a complex but monotonously decreasing function of the number of active components. For the resonators, the following physical limits of the lifetime exist.

Caesium beam tubes: saturation of the getter materials absorbing the spent caesium vapour.

Rubidium vapour cells: Ageing and blackening of the spectral lamp (easily replaceable in good designs).

Crystals: Eventually fatigue in the crystal mount (excessive ageing observed in some cases).

The large variety of designs and the still continuing development of improved devices make it appear unwise to state more precise figures here. The few remarks made above serve to illustrate the problem of reliability inherent to the attempt to use stable oscillators as clocks. More information on the problem of clock operation is given in Chapter 4.

3.4. FREQUENCY MULTIPLICATION AND DIVISION

3.4.1. Frequency multipliers

The principle of frequency multiplication, also called harmonic generation, is based on the Fourier series representation of periodic functions of time. Starting with a sinusoidal signal (represented either as a voltage or a current), applied to a nonlinear circuit element, we obtain a signal which still has the period of the original sinusoid but no longer a sinusoidal waveform. This resulting signal can be represented as a Fourier series composed of sinusoidal terms with angular frequencies ω (fundamental), 2ω (second harmonic), $n\omega$ (nth harmonic), the amplitude of the various terms depending on the nonlinear transfer characteristics producing the distorted waveform. The desired harmonic term is then extracted by a frequency selective filter and amplified, if necessary and possible. All harmonic generators operate according to this basic principle which is very simple indeed (Fig. 3.14).

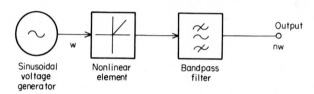

FIG. 3.14. Principle of harmonic generator.

Practical devices can be realized in different ways of which we will review only those which are of actual importance to measurement systems, leaving aside those techniques which are more appropriate for other fields, e.g. radio transmitters, etc.

Input signal: $U_0(t) = U_0 \cos \omega t$

period $T = 2\pi/\omega$

Nonlinear device

EXAMPLE 1: sequence of rectangular pulses (Fig. 3.15)

period T,   duration t_0

$$f(t) = \begin{cases} A & nT - t_0 < t < nT + t_0 \\ 0 & nT + t_0 < t < (n+1)T - t_0 \end{cases} \tag{3.23}$$

Duty cycle: $r = t_0/T$

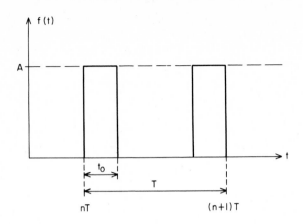

FIG. 3.15. Pulse sequence.

The (arbitrary) initial condition is made such that $f(t)$ is an even function $f(-t) \equiv f(t)$. Thus, the Fourier series contains only cosine terms which are all in phase.

$$f(t) = A_{av} + \sum_{n=1}^{n=\infty} C_n \cos n\omega t \qquad (3.24)$$

$$C_n = \frac{1}{\pi} \int_0^{2\pi} f(x) \cos nx \, dx \qquad (3.24a, b)$$

$$A_{av} = \frac{1}{T} \int_0^T f(t) \, dt.$$

The d.c. term A_{av} is not interesting for our further considerations, since we are only interested in the amplitude of the harmonic terms which is

$$C_n = 2Ar \frac{\sin nr}{nr} = 2A \frac{\sin nr}{n} \qquad (3.25)$$

with: $A_{av} = Ar$.

From this we see that with increasing order of harmonics, the amplitude decreases proportionally to $1/n$ and the power to $1/n^2$. This means that by using a *passive switching device having no internal amplification*, the power of the nth harmonic is theoretically limited to the fundamental input power divided by n^2. In practice, the major cause of loss by which the available harmonic power remains below the theoretical limit is that the switching mechanisms of the nonlinear element are not ideal. If we call P_0 the available fundamental and P_n the available harmonic power, we thus have

$$P_n < P_0/n^2 \tag{3.26}$$

for a non-amplifying passive switch harmonic generator.

Since for many applications, frequency multiplication to higher orders is needed, e.g. for caesium-beam frequency standards:

$$n \approx \frac{9200}{5} = 1840 \tag{3.27}$$

the frequency multiplication is done in several steps and the means for overcoming the harmonic power problem is amplification.

Phase noise, either inherent in the signal or generated by conversion of amplitude noise is an important phenomenon to be considered in the design and operation of frequency multipliers. It can be stated quite generally that the process of frequency multiplication always adds noise to the signal. The nonlinear element converts amplitude fluctuations into phase fluctuations. What happens in principle, can be easily understood from the example considered above.

To convert the sinusoidal input signal into the rectangular pulse sequence defined in (3.23) a switching operation has to be made at periodic intervals $t_0 + nT$, and at times when the input wave crosses a definite level. If the level of the signal fluctuates, the instants of switching fluctuate in time. If we have an ideal switch producing a step function and if we assume that the time fluctuations are very small compared to the period T, the components of the series (3.24) remain all in phase. The time fluctuations can thus be expressed as phase fluctuations according to Eq. (2.6):

$$\Delta\phi_n = n\omega\,\Delta t = n\,\Delta\phi_1 \tag{3.28}$$

i.e. the phase fluctuations expressed in radians, of the nth harmonic are equal to n times the phase fluctuations of the fundamental. This is again a theoretical minimum degradation assuming no additional phase fluctuations introduced by noise generated in the multiplier circuit elements. If multiplication to a high order n is done in several steps. the individual stages forming a "frequency multiplier chain", additional noise has the greatest effect in the lowest frequency stages. For these stages, the same careful design is required as for oscillator circuits if excessive signal deterioration due to phase noise is to be avoided.

The design of practical frequency multiplier circuits has to take into account the problems of power efficiency and phase noise outlined above. The need of filtering to extract the desired harmonic is common to all devices.

The most commonly used frequency multiplier types are the following:

passive diode harmonic generators;
transistor harmonic generators;
varactor frequency multipliers; and
step recovery diode harmonic generators.

All except the first one contain some amplification. A short review of the principle is given below.

Passive diode harmonic generators

In spite of its low power efficiency, especially for high order harmonics, the passive diode harmonic generator is a very practical device for many measurement applications up to very high frequencies. Metal point contact diodes have been used for harmonic generation in the far infrared region.[31] Its simplicity especially favours applications in laboratory experiments.

An approximate analysis of passive diode harmonic generation is useful to illustrate the main parameters involved in the design of diode and transistor harmonic generators.

Figure 3.16 shows the generation of current pulses in an idealized diode whose $i–u$ characteristic has been approximated by straight lines. The applied

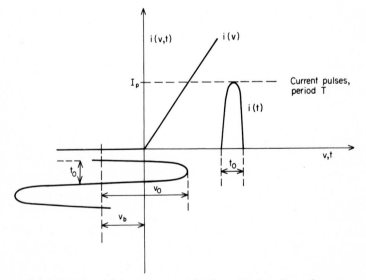

Fig. 3.16. Truncated sine pulse generation by an ideal nonlinear element.

voltage produces truncated sinusoidal current pulses once every fundamental period.

The peak current I_p can be calculated assuming the circuit configuration of Fig. 3.17. Here, we have a voltage source $u_0(t) = U_0 \sin \omega_1 t$ having an internal resistance R_0, connected through the diode to the primary winding of a tuned transformer having a turns' ratio $1:m$. (Stray inductance is neglected for sake of simplicity.)

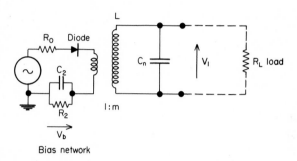

FIG. 3.17. Principle of diode harmonic generator.

L_n, C_n are resonant at the desired harmonic $n\omega_1$. We further assume that the input impedance of the transformer is small compared to R_0 at all frequencies. Then I_p is given by:

$$I_p = \frac{U_0 - U_b}{R_0 + R_d} \tag{3.29}$$

where R_d is the diode forward resistance defined by the slope of the conducting diode i–u characteristic (Fig. 3.16).

The bias voltage is obtained by means of the network $R_2 C_2$. The bypass capacitor C_2 represents a short circuit for all frequencies of interest. Thus:

$$U_b = I_{av} R_2 \tag{3.30}$$

I_{av} being the average value of the current pulses.

The duty cycle of the pulses is also obtained from Fig. 3.16:

$$r = \frac{t_0}{T} = \frac{1}{2} - \frac{1}{\pi} \arcsin \frac{U_b}{U_0} \tag{3.31}$$

The average value of the pulse train is given by:[39]

$$I_{av} = \frac{I_p}{\pi} \frac{\sin \pi r - \pi r \cos \pi r}{1 - \cos \pi r} \qquad (3.32)$$

and the Fourier coefficient for the nth harmonic by:

$$C_n = \frac{I_p}{\pi n} \frac{1}{1 - \cos \pi r} \left(\frac{\sin \pi r(n - 1)}{n - 1} - \frac{\sin \pi r(n + 1)}{n + 1} \right). \qquad (3.33)$$

Numerical analysis of (3.33) shows that the efficiency of harmonic generation falls off very rapidly with increasing harmonic order n. In Table 3.4 approximate values are indicated for the optimum duty cycle r_{opt} and the corresponding normalized Fourier coefficients $c_n = C_n/I_p$ for harmonic orders 2 to 5.

TABLE 3.4

$n \;=\;$	2	3	4	5
$r_{opt} =$ 0·333		0·21	0·167	0·14
$c_n \;=\;$ 0·276		0·184	0·139	0·111

In the design of a diode harmonic generator we are not only interested in its power efficiency but also in the quality of the generated output signal. Both can be estimated roughly by some simple calculations based on the equivalent circuit of Fig. 3.18. The inherent nonlinearity of the process makes an exact description rather involved. The approach given here must therefore not be regarded as being exact but rather as an illustration of the orders of magnitude involved. The current source I_1 represents the diode current pulses I_p, transformed by the transformer turns ratio m, thus

$$I_1 = I_p/m. \qquad (3.34)$$

R_1 is the diode and source resistance $(R_0 + R_d)$ transformed:

$$R_1 = \frac{1}{m^2}(R_0 + R_d) \qquad (3.35)$$

R_1 is not constant with time but present only during the conducting cycle t_0 (see Fig. 3.16). If R_1 is large compared to the load resistance R_L, its effect can be neglected. In this case, the assumption made before, i.e. a small input impedance of the transformer in Fig. 3.17 is also satisfied.

The allowed range of load resistance is limited by the minimum-Q of the resonant circuit required for filtering the desired harmonic. There are also limits to the physical realizability of inductors and capacitors with low loss. Impedance transforming networks are often required in order to present an optimum value of R_L to the terminals of the circuit in Fig. 3.17. A numerical example of a diode harmonic generator is given below:

EXAMPLE 2 (Diode harmonic generator):
Fundamental frequency: $\qquad\qquad\qquad\qquad f_1 = 5\,\text{MHz}$
Fifth harmonic, desired output frequency: $f_5 = 25\,\text{MHz}$
Fourier coefficient: $\qquad\qquad\qquad\qquad c_5 = 0\cdot111$
Duty cycle (see Table 3.3): $r = 0\cdot14 \qquad U_0 = 5\,\text{volt (peak)} \qquad R_0 = 50\,\text{Ohm}$
Diode: $R_d = 100\,\text{Ohm}$
Required bias voltage (from Eq. (3.31)):

$$\frac{U_b}{U_0} = \sin\pi(\tfrac{1}{2} - r) = 0\cdot904 \qquad U_b = 4\cdot52\ V$$

Peak current (from Eq. (3.29)):

$$I_p = \frac{0\cdot48}{150} = 3\cdot2\,\text{mA}$$

Average current (from Eq. (3.32)):

$$I_{av} = 0\cdot298\,\text{mA}$$

Bias resistor: $R_2 = U_b/I_{av} = 15\cdot2\,\text{kOhm}$
Bypass capacitor: $C_2 \sim 5\,\text{nF}$, so that $\omega_1 R_2 C_2 > 10^3$
Assumed transformer ratio: $m = 8$
Peak secondary current:

$$I_{1p} = \frac{I_p}{m} = 0\cdot4\,\text{mA}.$$

Transformed source resistance during conducting period:

$$R_1 = m^2(R_0 + R_d) = 9600\,\text{Ohm}$$

Assumed load resistor $R_L < R_1$: $R_L = 2000\,\text{Ohm}$
Effective load: R_1 parallel R_2: $1655\,\text{Ohm} = R'_L$
Peak fifth harmonic voltage:

$$U_5 = I_{1p} \times R_L \times C_5 = 0\cdot073\ V\ \text{peak}$$

RMS—fifth harmonic voltage:

$$U_{5\,\text{RMS}} = 0.052 \text{ V}$$

Power dissipated in 2000 Ohm load:

$$P_5 = \frac{U_{5\,\text{RMS}}^2}{R_L} = 1.35 \times 10^{-6} \text{ W}$$

Power available from source:

$$P_{10} = \frac{U_0^2}{8R_0} = 62.5 \text{ mW}$$

Power efficiency, including mismatch loss:

$$\eta = \frac{P_5}{P_{10}} = 2.16 \times 10^{-5} = -47 \text{ dB}$$

Resonant circuit parameters:
$$Q = 50 \text{ assumed with load}$$

$$\omega_5 L_5 = \frac{R'_L}{Q} = 33.1 \text{ Ohm}$$

$$L'_5 = 0.21 \text{ µH} \qquad C'_5 = 190 \text{ pF}$$

The residual amplitude modulation is determined by the circuit Q. The resonant circuit is driven to the computed peak value with every fifth cycle and the amplitude decays as a free damped oscillation according to the well known relation

$$A(t) = A_0 \exp\left(-\frac{\omega_n t}{2Q}\right)$$

with $Q = 50$, at the fourth cycle, $\omega_5 t = 8\pi$ and the amplitude has decayed to:

$$A(\omega_5 t = 8\pi) = 0.78\, A_0.$$

The amplitude of the output signal is thus modulated by a periodic exponential waveform. Sidebands at multiples of the fundamental frequency are generated. The amplitude and power of the sidebands can be calculated by harmonic analysis. If the output circuit is detuned from resonance, there is simultaneous amplitude and phase modulation. The sideband amplitude distribution is then no longer symmetric with respect to the carrier frequency.

As already stated, this example is rather an illustration of orders of magnitude than an exact design. In practice, the ideal diode characteristic is not realized. Especially at high frequencies, the finite switching time of the diode may cause further attenuation of the desired harmonic power.

Transistor harmonic generators

In transistor harmonic generators, the principle of pulse shaping by the nonlinear transfer characteristic of a correctly biased transistor (Class "C" operation) is very similar to that of the diode harmonic generator discussed above. However, superior performance is obtained due to the amplifying and impedance transforming properties of transistors. Bipolar and field effect transistors can be used within their respective capabilities and fairly predictable performance is obtained at output frequencies for which the transistor would perform well as an r.f. amplifier.

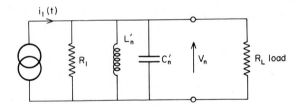

Fig. 3.18. Harmonic generator equivalent circuit.

With some types of bipolar transistors, the amplitude of the drive voltage is limited by the reverse breakdown voltage of the base-emitter junction. Otherwise, the equivalent circuit of Fig. 3.18 can be used to evaluate the output power of a transistor harmonic generator, the current source being that of the usual transistor equivalent circuit using admittance parameters, with $R_1 \approx 1/y_{22}$. Transistor harmonic generators can be designed for low order multiplication with much higher efficiency than that of the passive diode circuit described before. Usually, the output power is sufficient to drive another multiplier stage connected in cascade. High order multiplication factors can thus be obtained from low frequencies up through the VHF range, provided that the desired factor can be decomposed in small prime numbers, e.g. 2, 3 or 5.

The limit of practically useful frequency multiplication factors is mainly the level of phase noise which can be tolerated. In addition to the inherent phase noise multiplication mentioned earlier, there is also a phase noise contribution by the p–n junctions of the transistors. For almost all conventional devices this noise contribution at low Fourier frequencies is of the order of $S_\phi \approx 10^{-11\cdot2} f^{-1}$ (radian^2 Hz^{-1}).[40]

If excess gain is available, the phase noise contribution of the active element can be reduced by means of local negative feedback. Furthermore, some selected devices of certain types of transistors show less than the indicated phase noise for reasons which are not yet well understood. There is no established correlation between low noise figure (for additive white noise) and phase flicker noise, but there are indications of some relation between low frequency d.c. flicker noise and phase noise. Component selection by means of phase noise measurement techniques such as those described in Chapter 7 can be used if the highest possible performance is required, especially at the lowest frequencies in the system.

Varactor frequency multipliers

Varactor frequency multipliers are based on the nonlinear properties of reverse biased p–n junction diodes. The capacitance of a varactor diode depends on the applied reverse bias voltage U_b in the following way:

$$C(U_b) = \frac{C_0}{(1 + (U_b/\phi))^\gamma} \qquad (3.36)$$

where

$$\phi \sim 0{\cdot}1 \ \ldots 1 \text{ volt}$$

$$\gamma \sim 0{\cdot}33\ldots0{\cdot}5.$$

The potential ϕ depends on the properties of the semiconductor material and the exponent γ on the junction doping profile. C_0 is the junction capacitance at zero bias voltage and the relation is valid in the range between $U_b = 0$ and the reverse breakdown voltage of the diode.

The theory of varactor applications to parametric amplifiers and harmonic generators is described in detail in the literature.[41]

Development work in the early sixties had shown that the high efficiency predicted by the basic theory could be obtained only for low order (2 or 3) harmonic generators and it always remained difficult to achieve stable operation, especially for wide variations in drive level. Further development of VHF and UHF power transistors and the introduction of step recovery diodes have provided alternatives which are easier to design for good reliability.

For this reason, varactor frequency multipliers are now of historical interest rather than of practical importance, at least in the field of frequency measurement techniques.

Step recovery diode harmonic generators
The characteristic properties of step recovery diodes[42] were discovered during the development of varactor diodes. Throughout the experimental work on varactor harmonic generators it has been observed that improved efficiency could be obtained by driving the diode into forward conduction during part of the cycle. Further investigation of *p–n* junction properties has

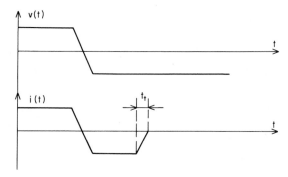

FIG. 3.19. Step recover diode voltage and current waveforms with resistive load.

then shown that, with a special *p–i–n* doping profile, leaving a narrow intrinsic region between the *p* and *n* region, it is possible to create a device which has the properties of an extremely fast, charge controlled switch. In the reverse biased condition, after removal of the conduction carriers, the capacitance is low and almost independent of the reverse bias voltage. In the forward conducting region the carrier density is high and thus also the capacitance. If the applied voltage is reversed from forward to reverse the current too is reversed and continues to flow in the inverse direction until the charge stored in the junction is removed. At this instant, the reverse current drops to zero in an extremely short time. Fig. 3.19 shows schematically this behaviour.[43] The amplitude of this current step is practically equal to the forward current if the duration of the current reversal is short compared to the carrier recombination lifetime in the junction.

The duration of the recovery step is called the "transition time" and can be of the order of 100 picoseconds or less. This property allows the generation of very sharp pulses with an energy content limited by the forward current intensity and the reverse breakdown voltage. A fairly exact theory of step recovery diode harmonic generation is described in Ref. 44. The main difference between the step recovery diode harmonic generator and the passive diode circuit described before is that in the former there is an efficient amplifying mechanism by charge storage and transfer. Therefore, the efficiency for higher order harmonic generation is not restricted by the $1/n^2$ limit but approaches the order of $1/n$. The main losses are due to carrier recombination, finite forward resistance and external circuit losses. The upper frequency limit is determined by the finite transition time.

For applications in measurements, two types of harmonic generators are useful, namely:

(a) broadband "comb" generators; and
(b) narrowband tuned harmonic generators.

The comb generators are used for frequency marker displays in spectrum analysers, swept-frequency transmission measurement systems and heterodyne converters for microwave frequency counters (see Section 5.3).

In a comb generator, the step recovery diode is used to generate a sequence of sharp pulses with a repetition rate equal to the input frequency. The carrier recombination lifetime puts a lower limit on the input frequency if a sinusoidal input signal is used, since the current reversal must be short compared to that lifetime, in order to prevent loss of stored charge due to carrier recombina-

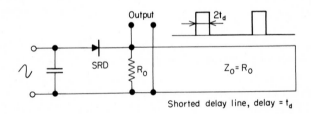

FIG. 3.20. Principle of series mode SRD (step recovery diode) pulse generator.

tion. Fig. 3.20 shows a simplified diagram of a series mode pulse generator using a short-circuited delay line for pulse shaping.[43] The duration of the pulses is equal to twice the propagation time of the delay line.

Tuned harmonic generators are designed to produce one particular multiple of the input frequency. Examples are given in Ref. 44.

3.4.2. Frequency dividers

Frequency division is in some way a complementary operation to frequency multiplication. A frequency division by a number n means that a determined phase of every nth cycle of the periodic input waveform is to be selected. This can be done in two basically different ways:

(a) synchronization; and
(b) counting.

Until a few years ago, before the introduction of cheap and reliable integrated logic circuits, we would have considered both methods as to be equally important. Now, the main interest has shifted to counting methods. Synchronization methods[43] use the possibilities of triggering a free-running relaxation oscillator by a pulse train derived from the input signal. The timing network (usually an RC network) is adjusted so that the free-running oscillator's period is just above n-times the period of the input pulses.

Almost any type of relaxation oscillator can be triggered in this way. The main problem with these devices is that the correct division ratio depends on the stability of the timing network elements as well as on the supply voltage. Except for very low division ratios, component ageing can eventually cause erratic or entirely false triggering. Near the limit but still operating correctly, these circuits also become very sensitive to external interference. The main advantage of synchronized relaxation oscillator dividers, namely low component count, has practically become extinct due to the advent of integrated circuit (IC) technology.

We therefore shall review only frequency division by counting and of these methods only the most important cases since abundant literature exists on digital and logic circuits.

The basic building block of a counting circuit is the flip–flop, a device which can assume either one of two stable logic states, depending on the logic levels at its input connections. Transitions between those stable states are initiated by a corresponding change of an input logic level. The simplest form of a divider, namely division by two, is obtained if the flip–flop state is changed upon every second input logic transition. Cascading a number n of such binary divider stages allows to divide the input frequency by a ratio of $N = 2^n$. Division by other ratios than binary numbers is obtained by inhibiting some logic transitions by means of gates. Therefore, the number n of flip–flops required is always such that the desired division ratio N is

$$N \leqq 2^n,$$

the equal sign applying to the case of binary division. Thus, at least in principle, any number of flip–flops can be arranged in cascade with appropriate gates to realize any desired division ratio. Practical limits exist due to the finite switching speed and delay introduced by the flip–flops and gates. During the fast and extensive evolution of IC technology since the early sixties, various families of IC technology have been developed which differ as to speed, power consumption and noise immunity. The high cost of initial IC design has led to some standardization in the form of basic building blocks which are produced in large quantities and therefore cheaply. In the older circuit design technique, cost was mainly determined by the number of active elements such as transistors and the number of required connections. With IC's, the complexity of the internal circuitry is less important as a cost factor inasmuch as it might limit the production yield. That this barrier is receding from year to year is shown by the more recent developments of large-scale integration resulting in the advent of entire small computers on a single silicon chip.

The basic J–K flip–flop[45] unit most commonly used in frequency dividers and counters is much more complex than its older discrete transistor or tube counterpart.[43] Other versions such as the R–S or D flip–flop[45] can also be used in special cases, however, the J–K flip–flop is the most suitable unit for universal application since it contains most of the otherwise externally required gating circuitry. Figure 3.21 shows the logic block diagram of a

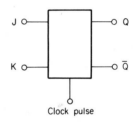

FIG. 3.21. *J–K* Flip–Flop.

J–K flip–flop. It has five terminals for connection, namely a clock pulse (CP) input, two input terminals designated by J and K which by application of logic levels serve to programme the change or no change of the logic state (and level) of the output terminal Q and its logic complement $\bar{Q}$ (if $Q = 0$, then $\bar{Q} = 1$ and vice versa). The basic rule of operations of a J–K flip–flop are the following:

(a) the timing of logic state change is determined by the clock pulse only;
(b) the J and K input terminals must have definite and constant levels when the clock pulse arrives:
(c) the output state changes unconditionally from the previous state only if both J and K terminals are high (logic 1); and
(d) all possible states are defined as functions of the J and K input levels by Table 3.5 (called the *characteristic table*).[45]

TABLE 3.5. *J–K* flip–flop characteristic table.

J	K	Q^{n+1}
0	0	Q^n
0	1	0
1	0	1
1	1	$\overline{Q^n}$

Table 3.5 shows what will be the state Q^{n+1} following the clock pulse when Q^n was the previous state at the Q-terminal, depending on the four possible combinations of logic levels at the inputs J and K. In plain language, this means that if both $J = K = 0$, the Q-level remains at its previous state. If $J = 0$, $K = 1$, the Q-level will be zero. If $J = 1$, $K = 0$, the Q-level will be one and if $J = 1, K = 1$, the Q-level will be the complement (will have changed) whatever its previous level.

For the purpose of programme synthesis, Table 3.6 (called the *excitation table*) can be useful. This table shows what J and K logic inputs are required in order to make the flip–flop respond in one of the four possible ways, i.e. two changes and two "no-changes". The word "any" in the right-hand columns means exactly that the desired response occurs irrespective of that level.

TABLE 3.6. *J–K* flip–flop excitation table.

Q^n	Q^{n+1}	J	K
0	0	0	any
0	1	1	any
1	0	any	1
1	1	any	0

The application of *J–K* flip–flops in frequency division is best illustrated by means of one typical example. Fig. 3.22 shows one of several possible configurations in which four flip–flops can be combined to produce a binary coded decimal (BCD) counter or decade counter or divider. This example shows the variety called synchronous counter (for other possibilities, see Refs. 43, 45), in which the clock inputs CP are all connected directly. Thus all flip–flops programmed to change state do so at the same time. The *J–K* inputs are connected to appropriate outputs through AND-Gates performing the Boolean equations given below each gate. The *J–K* inputs of flip–flop A are connected

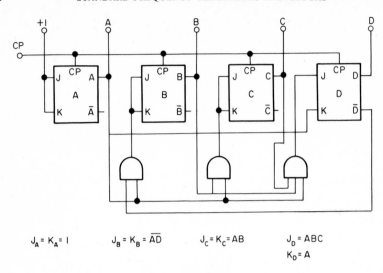

$$J_A = K_A = 1 \qquad J_B = K_B = \overline{AD} \qquad J_C = K_C = AB \qquad J_D = ABC$$
$$K_D = A$$

FIG. 3.22. BCD Synchronous counter.

to a constant logic 1 level (e.g. the positive power supply line). Step by step analysis of the sequence of possible states (see Ref. 45) leads to those listed in Table 3.7.

The same sequence of logic states is shown in Fig. 3.23 as a timing diagram. For use of this type of counter, the input frequency is applied at the CP input. Inspection of the timing diagram shows that signals of the output frequency, i.e. the input frequency divided by ten, are available at the terminals D as well

TABLE 3.7. Synchronous counter state table.

State, clock pulse No.	Logic states			
	D	C	B	A
0	0	0	0	0
1	0	0	0	1
2	0	0	1	0
3	0	0	1	1
4	0	1	0	0
5	0	1	0	1
6	0	1	1	0
7	0	1	1	1
8	1	0	0	0
9	1	0	0	1
10 = 0	0	0	0	0
etc.				

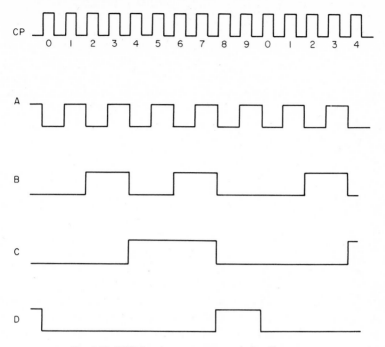

FIG. 3.23. BCD Synchronous counter timing diagram.

as C. It depends on the application of the output signal as to which terminal is to be preferred. If a sinusoidal output is to be obtained by means of bandpass filtering and amplification, the signal appearing at C is more useful because of its higher duty cycle. For driving further dividers either terminal can be used in principle. Inspection of Table 3.7 shows that the four terminals D, C, B, A represent the count numbers in binary form. The input signal waveform is shown in Fig. 3.23 as a symmetrical rectangular pulse sequence. If the original signal is a sinusoid as generated by an oscillator, well-known means of wave-form shaping[43] are to be applied in order to generate a clock pulse sequence with optimum rise time and duty cycle as required by the particular type of IC logic used.

BCD decade counters or dividers can be obtained in various configurations and technologies as IC's. Several versions are available which allow pre-setting to a wide range of division ratios. As the technology is still evolving, the reader is advised to consult the current literature and documents published by the semiconductor manufacturing companies.

The use of IC counters as frequency dividers is limited on the high fre-quency side by the achievable switching speed of the circuits. Toggling

D

frequencies in excess of 1 GHz have been recently achieved for simple binary dividers. More complex configurations as BCD or preset divider circuits must work at slower speed because of the delays involved in the gating circuitry.

Since frequency division is a timing process, phase fluctuations of the input signal are transmitted through the divider as time fluctuations. If we would neglect any time fluctuation (also called "timing jitter") added by the divider circuits, then the phase fluctuations expressed in radians at the output frequency $f_N = (1/N)f_1$ would be equal to

$$\Delta\phi_N = \frac{1}{N}\Delta\phi_1 \qquad (3.37)$$

Comparing this with Eq. (3.11), we may conclude that frequency division has, in a first approximation, the opposite effect on phase noise, compared to frequency multiplication, namely an improvement by the division ratio N. However, this is only true in the limit where the resulting output phase fluctuations are still caused by the input signal. The residual timing jitter generated by the logic circuitry itself remains as a lower limit. In this context, the noise immunity of the divider circuits is an important factor. False triggering by pulses arriving on another than the intended signal path can have disastrous consequences on the performance. Especially with low-voltage high-current devices as TTL and ECL logic, the power supply system including the layout on circuit boards must be carefully designed. The prevention of false counts is especially important in counting dividers used in clock systems because the continuity of the time scale is directly affected. Therefore, redundancy is used in applications where continuity is required.

One possible configuration is a triple parallel divider chain with the outputs combined in a "majority-voting" scheme and automatic reset of the faulty minority. In larger establishments, several independent divider systems having separate power supply systems are used.

3.5. FREQUENCY SYNTHESIS

In the two preceding sections we have reviewed frequency multipliers and dividers, i.e. devices which perform two basic arithmetic operations on frequencies. In many applications, not only for measurements but also for operational systems in the fields of telecommunications, radar and navigation, many signals of various frequencies are required, sometimes with high stability. Variable frequency oscillators often do not fulfill the stability requirements whereas crystal oscillators are tunable over a narrow range only. This gap can be filled by means of frequency synthesizers—devices which perform several or all of the four basic arithmetic operations on the frequency generated by a master oscillator (crystal or atomic). The field of frequency synthesis has gone through a rapid and extensive evolution during the last 25 years. This and the actual state of the art are described in a recently published book by V. F. Kroupa.[46] The study of this excellent text is recommended to all readers interested in frequency synthesis and its existence allows us to be very brief here.

The most primitive form of frequency synthesis is performed by a combination of a multiplier and a divider as shown in Fig. 3.24. If a programmed divider is used, some limited means of output frequency variation are available. More flexible arrangements are possible through the introduction of *addition* or *subtraction of frequencies*.

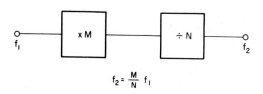

FIG. 3.24. Simple frequency synthesizer.

This is another fundamental operation on signals and can be obtained by means of amplitude modulators or mixers. Amplitude modulation is a large-signal nonlinear operation which on sinusoidal signals is performed basically by means of generating the product of two voltages. Given: $U_1(t) = U_1 \cos \omega_1 t$; $U_2(t) = U_2 \cos \omega_2 t$ the result $f(t)$ is:

$$f(t) = U_1 U_2 \cos \omega_1 t \cos \omega_2 t$$

$$= \frac{U_1 U_2}{2} (\cos (\omega_1 + \omega_2) t + \cos (\omega_1 - \omega_2) t). \qquad (3.38)$$

by applying an elementary trigonometrical formula. The result contains the sum of two signals having sum and difference frequencies respectively. These signals could then be separated by appropriate filters. In practice, however, this ideal condition is never realized.

A nonlinear element containing a square law term on which the sum $U_1 + U_2$ is applied in some way would produce a term $(U_1 + U_2)^2 = U_1^2 + 2U_1 U_2 + U_2^2$ but also many other terms called intermodulation products. In Ref. 46 a concise but general approach on nonlinear modulation theory is given, including references to earlier and more detailed work in this

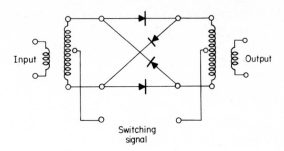

FIG. 3.25. Double balanced ring modulator.

field. The non-ideal performance of mixers is one of the main problems in the design of frequency synthesizers for high spectral purity of the output signal as undesired intermodulation products may show up as spurious components of the output signal. The closest approximation to the ideal mixer of Eq.(3.38) is obtained by means of double balanced switching modulators of which the ring modulator shown in Fig. 3.25 is a well-known example. As described in

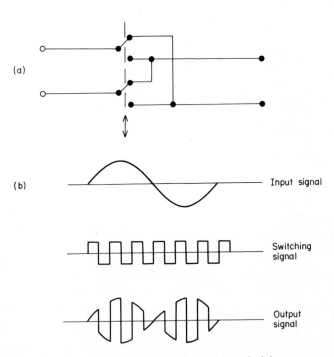

FIG. 3.26. Double balanced modulator principle.

Ref. 46, this circuit operates like a phase-inverting switch according to Fig. 3.26. Ideally, such a switch produces the following result:

$$u_1(t) = U_1 \cos \omega_1 t \quad \text{input signal}$$

$u_2(t)$ square-wave switching function, i.e.

$$u_2(t) = \frac{4}{\pi} \frac{(-1)^{k-1}}{2k-1} \sum_{k=1}^{\infty} \cos k\omega_2 t$$

$$u_3(t) = U_1(t) U_2(t) \qquad U_3 = \frac{4U_1}{\pi}$$

$$u_3(t) = U_3 \cos \omega_1 t \frac{(-1)^{k-1}}{2k-1} \sum_{k=1}^{\infty} \cos k\omega_2 t$$

$$= U_3 \left[\cos \omega_1 t \cos \omega_2 t - \tfrac{1}{3} \cos 3\omega_1 t \cos \omega_2 t + \ldots - \ldots \right]$$

$$= U_3 \frac{(-1)^{k-1}}{2k-1} \sum_{k=1}^{\infty} \left[\cos (k\omega_2 + \omega_1) t + \cos (k\omega_2 - \omega_1) t \right] \qquad (3.39)$$

i.e. the input signal and the switching signal frequencies and their harmonics do not appear in the result but only sum and difference frequencies between the input signal and the switching signal fundamental and odd harmonic frequencies. In practice, other intermodulation products are not entirely suppressed but nevertheless strongly attenuated. Double balanced mixers using Schottky–Barrier diodes and subminiature toroidal core input and output transformers are currently available from various manufacturers as standard circuit elements either for printed circuit board mounting or as shielded units with coaxial connectors. Some of these mixers have extremely low phase noise and are therefore used in phase-noise measurement setups (see Chapter 7).

Two different approaches can be distinguished in the practical design of frequency synthesizers:

(a) direct synthesis, comprising only multipliers, dividers, mixers and required filters; and
(b) indirect synthesis, comprising oscillators which are synchronized by means of phase-locked loops.

Elementary loops for *direct synthesis* are shown in Fig. 3.27, including one multiplier, one divider and one mixer.

There are some limits to the design of these loops (Ref. 46, p. 124) which can be summarized as follows:

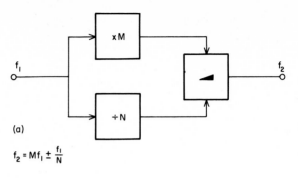

(a)

$$f_2 = Mf_1 \pm \frac{f_1}{N}$$

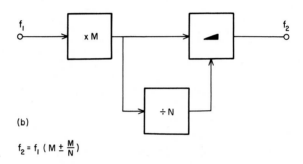

(b)

$$f_2 = f_1 \left(M \pm \frac{M}{N} \right)$$

FIG. 3.27. Elementary direct synthesis loops.

The multiplication factor M should be restricted to products of small prime numbers (2, 3, perhaps 5).

The ratio $|q| > 0$ of frequencies added or subtracted in the mixer must be kept within reasonable limits.

A rule similar to the second one above has been stated by D. G. Meyer.[47] In addition or subtraction of the form

$$f_2 = f_0 \pm f_1 \quad \text{where } f_1 < f_0$$

one should have

$$\frac{f_2}{f_1} < 10. \tag{3.40}$$

Observation of this rule makes the design of the output filter easier for attenuating spurious intermodulation products. More elaborate synthesizers containing several interconnected loops must be designed with careful planning of the frequencies and signal levels involved, especially with regard to the desired spectral purity of the output signal. In direct synthesis, the output signal phase is directly connected to that of the input signal. If the latter is disconnected, no output signals is generated. With programmed dividers, the output frequency can be switched to a new value in a very short time and with a minimum of transient phase variations.

Indirect synthesis by means of phase-locked loops is illustrated by the example of Fig. 3.28. For sake of simplicity, only frequency dividers are

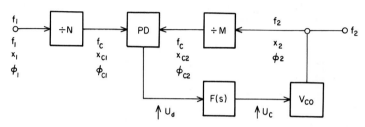

FIG. 3.28. Example of indirect synthesis by means of a phase-locked loop.

included in this example. The reference frequency f_1 and the output frequency f_2 are both divided by entire numbers N and M respectively, to a common submultiple frequency f_c:

$$\frac{f_1}{N} = \frac{f_2}{M} = f_c \tag{3.41}$$

Thus

$$f_2 = \frac{M}{N} f_1 \tag{3.42}$$

The phase detector (PD) generates a voltage proportional to the phase difference of the signals at f_c. This voltage controls, through a loop filter (LF), the frequency of the voltage controlled oscillator (VCO). In the steady state, the frequency error of the VCO is zero. If the phase fluctuations of the reference f_1 and of the VCO are small compared to 1 radian, small signal linear theory can be applied and the behaviour of the loop described by well-known transfer functions of classical control theory. The transient phenomena occurring during acquisition of the lock are nonlinear and more difficult to describe, except in very simple cases. Phase-locked loop theory applied to frequency synthesis is discussed in detail in Ref. 46, based mostly on the general work of Gardner.[48] Therefore, only a short description of the small signal behaviour of the locked synthesizer loop will be given here. In the usual approach, as shown in the cited references, the loop equations are established as relations between input and output *phase angles* (expressed in radians or degrees), respectively as equations between the Laplace transforms of these quantities.

As an alternative to the usual phase-angle formalism, we show here the corresponding relations between the *phase-times* of input and output signals. This phase-time formalism leads to slightly simpler expressions than the phase-angle formalism and is consistent with the basic theory given in Chapter 2. The variables involved are given in Fig. 3.28.

Phase-time formalism

The *phase detector* generates a d.c. voltage proportional to the phase-time difference between the signals at the common divided frequency f_c:

$$U_d(t) = A f_c (x_{c1} - x_{c2}). \tag{3.43}$$

The dimension of A is in volts: $[A] = [V]^*$. If a linear ramp phase detector is used, A is equal to the operation range in volts of the detector and (3.43) is valid for $(x_{c1} - x_{c2}) \leqslant 1/f_c$, i.e. for the small signal condition throughout the operating range of the detector.

Except for the introduction of a constant time delay, the frequency divider does not affect the phase-times of the signals.

Considering the arbitrary initial conditions of the display frequency dividers, we have

$$x_{c1} = x_1 + n T_1 \tag{3.44a}$$

$$x_{c2} = x_2 + m T_2, \tag{3.44b}$$

where n and m are arbitrary integers defining the initialization states of the dividers. If we disregard this initialization, Eq. (3.43) can be rewritten in the form:

$$U_d(t) = A f_c (x_1 - x_2) \tag{3.45a}$$

and between the corresponding Laplace transforms:

$$U_d(s) = A f_c (X_1 - X_2). \tag{3.45b}$$

* Square brackets mean "dimension of" the quantity written in between.

The *loop filter* action is described by its general transfer function $F(s)$:

$$U_c(s) = F(s) U_d(s) \tag{3.46}$$

relating the Laplace transforms of U_d and U_c.

The *voltage controlled oscillator* is described by the equation relating the control voltage U_c to the oscillator's normalized frequency offset y_2:

$$y_2 = \frac{dx_2}{dt} = KU_c + y_{20} \tag{3.47}$$

$[K] = [V^{-1}]$ is the voltage control coefficient and y_{20} is the initial offset of the free-running oscillator. For practical oscillators (3.47) is usually valid only over a restricted operating range since most frequency control elements (e.g. varactors) are not linear. Equation (3.47) does not include any spontaneous phase fluctuations of the VCO. In order to take care of this important parameter, we rewrite (3.47) in integral form,

$$x_2 = K \int_0^t U_c(t)\, dt + \delta x_2 + y_{20} t \tag{3.48}$$

introducing δx_2 as a small random variable, $\delta x_2 \ll 1/f_2$, representing the phase-time fluctuations of the free-running oscillator. The Laplace transform of (3.48) is

$$X_2 = \frac{KU_c}{s} + X_{20} + \frac{y_{20}}{s^2} \tag{3.49}$$

provided $X_{20} = \int_0^\infty e^{-st}\, \delta x_2(t)\, dt$ exists.

Defining the overall d.c.-loop gain by $K_x = KAf_c$ and substituting, we obtain

$$X_2 = H_x(s) X_1 + (1 - H_x(s)) X_{20} + \frac{y_{20}}{s^2}(1 - H_x(s)) \tag{3.50}$$

with the input–output transfer function of the system:

$$H_x(s) = \frac{K_x F(s)}{s + K_x F(s)} \tag{3.51}$$

and the spectral density S_{x2} of the output phase-time fluctuations is related to the spectral densities S_{x1} of the reference and S_{x20} of the VCO by the formula:

$$S_{x2}(f) = |H_x(2i\pi f)|^2 S_{x1}(f) + |1 - H_x(2i\pi f)|^2 S_{x20}(f) \qquad (3.52)$$

for all $f > 0$.

The third term of (3.50) does not contribute to the result since y_{20} is a constant. It is worth mentioning that in the phase-time formalism, the divider ratios N and M do not appear. However, they are implicitly present in Eq. (3.43) defining the phase-detector action, namely that the gain of the phase-detector is proportional to f_c. The lower the comparison frequency, the lower the loop gain!

We now repeat the same example using the classical phase-angle formalism as in Ref. 46*:

Phase-angle formalism
Phase detector. Condition: $\phi_{c1} - \phi_{c2} < 2\pi$

$$U_d(s) = K_d(\Phi_{c1} - \Phi_{c2}) \qquad (3.53)$$

compared to (3.45b) we have: $K_d = A/2\pi$ and $[K_d] = [V/\text{radian}]$. Here, however, the phase *angles* are changed by the frequency dividers:

$$\phi_{c1} = \frac{\phi_1}{N}, \qquad \phi_{c2} = \frac{\phi_2}{M},$$

$$\Phi_{c1} = \frac{\Phi_1}{N}, \qquad \Phi_{c2} = \frac{\Phi_2}{M},$$

thus

$$U_d(s) = K_d\left(\frac{\Phi_1}{N} - \frac{\Phi_2}{M}\right) \qquad (3.54)$$

Loop filter: $U_c = F(s) U_d$ *as before in* (3.46).

Voltage controlled oscillator. By analogy to (3.48), we have

$$\phi_2 = K_2 \int_0^t u_c \, dt + \delta\phi_2 + \omega_{20}t \qquad (3.55)$$

---

* Ref. 46 does not treat exactly this example but very similar ones.

where the voltage control coefficient is now:

$$K_2 = K\omega_2 \quad \text{and} \quad [K_2] = [\text{radian/s} \cdot \times \text{V}], \quad \text{and} \quad \omega_{20}$$

the free running oscillator angular frequency. We obtain after transformation:

$$\Phi_2(s) = \frac{K_2 U_c}{s} + \Phi_{20} + \frac{\omega_{20}}{s^2} \tag{3.56}$$

$\delta\phi_{20}$ and its transform Φ_{20} are again introduced as the small spontaneous phase fluctuations (phase jitter, phase noise) of the VCO, and the existence of Φ_{20} assumed as before and as in Ref. 46.

The d.c. loop gain is now defined as $K_0 = K_2 K_d$ and we obtain by substitution:

$$\Phi_2 = \frac{M}{N} H_2(s)\,\Phi_1 + (1 - H_2(s))\,\Phi_{20} + (1 - H_2(s))\frac{\omega_{20}}{s^2} \tag{3.57}$$

where

$$H_2(s) = \frac{K_0 F(s)}{Ms + K_0 F(s)} \tag{3.58}$$

The relation between the output signal phase-noise spectral density $S_{\phi 2}$ and the spectral densities $S\phi_1$ and $S\phi_{20}$ of the reference and VCO phases respectively is:

$$S_{\phi 2} = \left(\frac{M}{N}\right)^2 |H_2(2\pi i f)|^2 \, S_{\phi 1} + |1 - H_2(2\pi i f)|^2 \, S_{\phi 20} \tag{3.59}$$

In this case the divider ratios M and N do appear in the equations.

It is easy to show that the two formalisms are entirely equivalent.* The preference to be given to either phase-time or phase-angle is not a fundamental question but rather a matter of taste.

The use of the phase-time formalism seems nevertheless to be more convenient for those applications in which problems of timing jitter arise.

---

* Comparing the definitions of $K_x = KAf_c$ and $K_0 = K\omega 2A/2$, we have $K_o = MK_x$. Thus $H_x(s)$ and $H_2(s)$ are identical. By remembering from Eqs (2.19) and (2.20), we have $S_{\phi 2} = \omega_2^2 S_{x2}$, $S_{\phi 1} = \omega_1^2 S_{x1}$, $S_{x20} = \omega_2^2 S_{\phi 20}$ and the equivalence of 3.52 and 3.59 follows directly by substitution.

This suggests its use in the analysis of timing and synchronization problems in digital systems. As pointed out by G. Becker (see Ref. 4 of Chapter 2), the concept of phase angle is closely associated with sinusoidal waveforms and its meaning becomes somewhat obscure if applied to pulse waveforms, whereas the concept of timing is more general.

The actual performance of a phase-locked loop frequency synthesizer depends on the numerical parameters of the systems. The loop filter transfer function $F(s)$ is usually either of the form

$$F(s) = \frac{1}{1 + sT_1} \quad \text{(single pole RC filter)} \tag{3.60}$$

or

$$F(s) = \frac{1 + sT_2}{1 + sT_1} \quad \text{(lag-lead filter)} \tag{3.61}$$

and the system analysis is performed using well-known servo system analysis techniques.[46]

For measurement purposes, the indirect synthesis methods have the advantage that the phase-locked loop acts as a filter for part of the reference signal noise, especially at higher Fourier frequencies. This is easily seen from (3.52) and (3.59) in which the second term becomes dominant at high values of f. The first term dominates at very low values of f, allowing the long term stability of the reference to be transferred to the synthesized output frequency. An important further constraint on the loop filter $F(s)$ and on the choice of f_c is sufficient attenuation of residual components of f_c in the control voltage. Otherwise the output signal will show spurious FM with the comparison frequency f_c.

In commercial instruments, synthesizer circuits are arranged as decade modules with the output signals combined to allow overall frequency ranges extending from a few kHz to several hundred MHz in steps of, for example, 1 Hz, 0·1 Hz or less.

Both direct and indirect synthesis methods are used with similar performance. In most recent instruments, indirect synthesis has allowed slightly lower noise levels to be obtained.

Improved design of phase-lock loops also has resulted in much faster switching speeds, almost approaching that of direct synthesizers.[49]

Applications of frequency synthesizers are growing very rapidly, especially due to the possibility of rapid frequency control by computer command, in automatic measurement setups, frequency agile radiocommunications systems, etc.

REFERENCES

1. E. A. Gerber and R. A. Sykes. Quartz crystal units and oscillators. *In* "Time and Frequency: Theory and Fundamentals" (B. E. Blair, Ed.), NBS Monograph 140. US Govt. Printing Office, Washington, DC. May 1974.
2. A. W. Warner. High frequency crystal units for primary frequency standards. *Proc. IRE*, **40**, 1030–1033 (1952).
3. E. Hafner. The piezoelectric crystal unit—Definitions and methods of measurements. *Proc. IEEE*, **57** (2), 179–201 (1969).
4. W. A. Edson. Noise in oscillators. *Proc. IRE*, **48** (8), 1454–1466 (1960).
5. E. Hafner. The effects of noise in oscillators. *Proc. IEEE*, **54** (2), (1966).
6. N. F. Ramsey. History of atomic and molecular frequency control of frequency and time. *In* "Proc. 25th Ann. Symp. on Frequency Control", pp. 46–51. US Army Electronics Command, Ft. Monmouth, NJ, April 1971.
7. R. E. Beehler, Cesium atomic beam frequency standards: A survey of laboratory standards development from 1949 to 1971. *In* "Proc. 25th Ann. Symp. on Frequency Control". pp. 297–308. US Army Electronics Command, Ft. Monmouth, NJ, April 1971.
8. "Comptes rendus de la treizième Conférence Générale des Poids et Mesures" (Paris, October 1967), p. 103. Gauthier-Villars, Paris, 1968.
9. K. Shimoda, T. C. Wang and C. H. Townes. Further aspects of the theory of the maser. *Phys. Rev.* **102**, 1308–1321 (1956).
10. J. H. Holloway and R. F. Lacey. Factors which limit the accuracy of cesium atomic beam frequency standards. *In* "Proc. Int. Conf. on Chronometry" (Lausanne, Switzerland), pp. 317–331. June 1964.
11. J. Viennet, C. Audoin and M. Desaintfuscien. Discussion of cavity pulling in passive frequency standards. *In* Proc. 25th Ann. Symp. on Frequency Control, pp. 337–342. U.S. Army Electronics Command, Ft. Monmouth, NJ, April 1971.
12. R. F. Lacey, A. L. Helgesson and J. H. Holloway. Short-term stability of passive atomic frequency standards. *Proc. IEEE*, **54**, 170–176 (1966).
13. R. L. Barger and J. L. Hall. Pressure shift and broadening of methane line at $3\cdot39\,\mu$ studied by laser-saturated molecular absorption. *Phys. Rev. Lett.* **22** 4–8 (1969).
14. H. Hellwig, H. E. Bell, P. Kartaschoff and J. C. Bergquist. Frequency stability of methane-stabilized He–Ne lasers. *J. Appl. Phys.* (January 1972).
15. A. G. Mungall. The second order Doppler shift in cesium beam atomic frequency standards. *Metrologia*, **7**, 49–56 (1971).
16. P. L. Bender, E. C. Beaty and A. R. Chi. Optical detection of narrow Rb^{87} hyperfine absorption lines. *Phys. Rev. Lett.* **1**, 311–313 (1958).
17. E. N. Fortson, F. C. Major and H. G. Dehmelt. Ultrahigh resolution $F = 0$, $\pm1(He^3)^+$ HFS spectra by an ion storage collision technique. *Phys. Rev. Lett.* **16**, 221–225 (1966).
18. N. F. Ramsey. "Molecular Beams". Clarendon Press, Oxford, 1956.
19. H. G. Dehmelt. Parametric resonance reorientation of atoms and ions aligned by electron impact. *Phys. Rev.* (*Lett.*), **103**, 1125–1126 (1956).
20. D. Kleppner, N. F. Ramsey and P. Fjelstad. Broken atomic beam resonance experiment. *Phys. Rev. Lett.* **1**, 232–233 (1958).
21. H. Hellwig. Private communication.

22. P. Davidovits and R. Novick. The optically pumped rubidium maser. *Proc. IEEE*, **54**, 155–170 (1966).
23. M. M. Zepler, T. J. Bennett, G. T. Norton and R. E. Hayes. Miniaturized rapid warm-up, rubidium frequency source. *In* "Proc. 25th Ann. Symp. on Frequency Control", U.S. Army Electronics Command, Ft. Monmouth, NJ, pp. 331–336. April, 1971.
24. P. H. Lee and M. L. Skolnick. Saturated neon absorption inside a 6238-Å laser. *Appl. Phys. Lett.* **10**, 303–305 (1967).
25. M. W. Levine and R. F. C. Vessot. Hydrogen-maser time and frequency standard at Agassiz Observatory. *Radio Science*, **5** (10), 1287–1292 (1970).
26. C. Finnie, R. Sydnor and A. Sward. Hydrogen maser frequency standards. *In* "Proc. 25th Ann. Symp. on Frequency Control", pp. 348–351. U.S. Army Electronics Command, Ft. Monmouth, N.J., April 1971.
27. D. J. Glaze. Improvements in atomic cesium beam frequency standards at the National Bureau of Standards. *IEEE Trans. Instrum. Meas.* **IM-19**, 156–160 (1970).
28. G. Becker, B. Fischer, G. Kramer, and E. K. Müller. Neuentwicklung einer Cäsium-Strahl Apparatur als primäres Zeit und Frequenz-Normal an der Physikalisch-Technische Bundesanstalt. *PTB-Mitteilungen*, **79**, 77–80 (1969).
29. H. Hellwig, R. F. C. Vessot, M. W. Levine, P. W. Zitzewitz, D. W. Allan and D. J. Glaze. Measurement of the unperturbed hydrogen hyperfine transition frequency. *IEEE Trans. Instrum. Meas.* **IM-19**, 200–209 (1970).
30. A. P. Kolchenko, S. G. Rautian and R. I. Sokolovskii. Interaction of an atom with a strong electromagnetic field with the recoil effect taken into consideration. *Sov. Phys.–JETP*, **28**, 986–990 (1969).
31. K. M. Evenson, J. S. Wells, F. R. Petersen, B. L. Danielson and G. W. Day. Accurate frequencies of molecular transitions used in laser stabilization: The $3\cdot39\,\mu m$ transitions in CH_4 and the $9\cdot33$ and $10\cdot18\,\mu m$ transitions in CO_2. *Appl. Phys. Lett.* **22**, 192 (1973).
32. H. E. Peters and D. B. Percival. NASA hydrogen maser accuracy and stability in relation to world standards. *In* "Proc. 4th Ann. PTTI Planning Meeting" (NASA Goddard Space Flight Center, Greenbelt, Md. 14–16 November 1972). NASA Doc. No. X-814-73-72.
33. H. Brandenberger, F. Hadorn, D. Halford and J. H. Shoaf. High quality quartz crystal oscillators: Frequency domain and time domain stability". *In* "Proc. 25th Ann. Symp. on Frequency Control," pp. 226–230, U.S. Army Electronics Command, Ft. Monmouth, NJ, April 1971.
34. D. H. Throne. A report on the performance characteristics of a new rubidium vapor frequency standard. *In* "Proc. 23rd Ann. Symp. on Frequency Control." pp. 274–278. U.S. Army Electronics Command, Ft. Monmouth, NJ, May 1969.
35. R. Hyatt, D. Throne, L. S. Cutler, J. H. Holloway and L. F. Mueller. Performance of newly developed cesium beam tubes and standards. *In* "Proc. 25th Ann. Symp. on Frequency Control." pp. 313–324. U.S. Army Electronics Command, Ft. Monmouth, NJ, April 1971.
36. G. M. R. Winkler, R. G. Hall and D. B. Percival. The U.S. Naval Observatory clock time reference and the performance of a sample of atomic clocks. *Metrologia*, **6**, 126–134 (1970). See also D. W. Allan and J. E. Gray. Comments on the October 1970 *Metrologia* paper, The U.S. Naval Observatory clock time reference and the performance of a sample of atomic clocks. *Metrologia*, 7, 79–82 (1971).
37. D. W. Allan, D. J. Glaze, H. E. Machlan, A. E. Wainwright, H. Hellwig, J. A. Barnes and J. E. Gray. "Performance, modelling and simulation of some cesium

beam clocks", *In* "Proc. 27th Ann. Symp. on Frequency Control". US Army Electronics Command, Ft. Monmouth, NJ, 1973.

38. D. J. Glaze, H. Hellwig, D. W. Allan, S. Jarvis Jr, A. E. Wainwright and H. E. Bell. Accuracy evaluation and stability of the NBS primary frequency standards. *IEEE Trans. Instr. Meas.* **IM-23** (4), 489–501 (1974).

39. ITT. "Reference Data for Radio Engineers", 5th Ed, pp. 42–44. Howard W. Sams Co. Inc., New York, March 1969.

40. D. Halford, A. E. Wainwright and J. A. Barnes. Flicker noise of phase in RF amplifiers and frequency multipliers: Characterization, cause and cure (Summary). *In* "Proc. 22nd Ann. Symp. on Frequency Control", pp. 340–341. US Army Electronics Command, Ft. Monmouth, NJ, April 1968.

41. P. Penfield Jr. and R. P. Rafuse. "Varactor Applications". The MIT Press, Cambridge, Mass., 1962.

42. J. Moll, S. Krakauer and R. Shen. P–N junction charge storage diodes. *Proc. IRE*, **50** (1), 43–53 (1962).

43. J. Millman and J. Taub. "Pulse, Digital and Switching Waveforms". McGraw-Hill, New York, 1965.

44. S. A. Hamilton and R. D. Hall. "Shunt Mode Harmonic Generation using Step Recovery Diodes. *Microwave Journal*, **10** (4), 69–78 (1967).

45. W. E. Wicks. "Logic Design with Integrated Circuits". John Wiley, New York, 1968. See also data and applications literature published by semiconductor manufacturers.

46. V. F. Kroupa. "Frequency Synthesis". Charles Griffin, London, 1973.

47. D. G. Meyer. An ultra low noise direct frequency synthesizer. *In* "Proc. 24th Ann. Symp. on Frequency Control", pp. 209–232. US Army Electronics Command, Ft. Monmouth, NJ, 1970.

48. F. M. Gardner. "Phaselock Techniques". John Wiley, New York, 1966.

49. W. F. Byers, K. W. Craft and G. H. Lohrer. A 500 MHz low noise general purpose frequency synthesizer. *In* "Proceedings of the 27th Annual Symposium on Frequency Control", pp. 180–190. US Army Electronics Command, Ft. Monmouth, NJ, 1973.

50. L. Essen and J. V. L. Parry. The caesium resonator as a standard of frequency and time, *Phil. Trans. Roy. Soc.* **973** (250), 45–69 (1957).

4

Time Scales

4.1. INTRODUCTION

In this chapter we review some concepts of time scales, based on the results of the introduction of highly accurate atomic frequency standards. We shall, therefore, not deal at all with philosophical concepts of time, and we will also leave aside considerations of gravitation and relativity theories, subjects which are well covered in the general scientific literature.* The main subject of this chapter is *physical time scales generated by clocks*, especially in the form of a running display or sequence of signals which are used in practice for the *dating of events* and for the control of various kinds of *time-ordered systems*.

The traditional way of timekeeping which is based on astronomical observations, geodesy, celestial mechanics and geophysics constitutes by itself a vast field of scientific activity and knowledge. It is, however, entirely outside the scope of this book which deals only with physical measurements, instruments and techniques. The following section will therefore give only a very brief account of the traditional time scales, just enough to suggest that the relationship between old and new ways of timekeeping is mainly historical.

4.2. CLOCKS AND TIME SCALES

Let us start by stating some important and basic facts: There is no easily available absolute time scale to be found in nature. An ideal device generating a perfect time scale is a purely theoretical construction. All real time scales

* See also Chapter 1.

based either on astronomical observations or on physical instruments are imperfect approximations to that theoretical concept.

The concept of a clock in a general sense, i.e. also including the rotating Earth and the movement of celestial bodies involves three main parts and functions, namely:

1. A periodic movement which can be observed.
2. The continuous counting of the periods.
3. The display of the registered count.

There is of course a large variety of possibilities to perform the three basic functions mentioned above. Whatever the means used to do this, the result will be a particular time scale defined by that particular clock system. That is, each clock defines its own time scale.

Among the periodic physical movements* used in time-keeping, we can distinguish three types leading to practical time scales:

A. *The free spinning rotor*
Example: the Earth.

The rotation of the Earth as observed by astronomical means leads to *Universal Time* (UT).

B. *Keplerian movement*
The movement of a satellite body around a central body, based on gravitation.

Examples: The revolution of the Earth around the Sun, of the Moon around the Earth, etc., which are the basis of *Ephemeris Time*.

C. *Harmonic oscillations*
Most kinds of mechanical or electrical oscillations* belong to this class, including the oscillations of an electromagnetic wave (photon) emitted or absorbed by a quantum-mechanical system. From this viewpoint, we can therefore point out as examples almost all practical precision oscillators and clocks:

    pendulum;
    balance wheel with hairspring;
    tuning fork;
    quartz crystal;
    atomic resonators.

* Nonperiodic movements are or have been used for time-keeping: Flow of water or sand, burning candles, radioactive decay, etc. However, none of these devices has yet been used successfully for the design of a precision timekeeper.

* Periodic but not harmonic (relaxation) oscillations are used in less precise instruments. The oldest example is the foliot used in early clocks. In modern electronics, astable multi-vibrators are widely used in simple timing circuits.

The types A and B of periodic movements are the basis for the concepts of *Astronomical Time* and *Astronomical Time Scales*.

The harmonic oscillations of type C lead to the great variety of devices known as clocks in the popular sense. Among these clocks, we distinguish the *Atomic Clocks* as being the most accurate and this leads us to the concept of *Atomic Time* and *Atomic Time Scales*.

Starting from any of the periodic movements mentioned above to establish a time scale by observation, continuous counting, registering and displaying, there are two main actions which have to be undertaken in order to obtain a time scale, namely:

1. The *period* or its inverse, the *frequency* of the basic oscillation has to be *measured*, *adjusted* or *defined*.
2. The *origin* of the time scale has to be *defined*.

The first action consists in establishing a unit of time interval but is not sufficient yet to define a time scale. Only the choice or definition of an origin from which we start counting the periods, completes the task. In practice, both actions require *conventions* to be agreed upon.

The origins of all practically used time scales are in fact based on conventions obtained by international agreement among the interested parties.[1,2]

Whatever type of phenomenon is used to establish a time scale, the most important requirement is its *uniformity*. Uniformity means that the intervals between the scale marks are equal, i.e. constant *period* and *frequency* of the basic oscillation.

Time Scale Uniformity and *Frequency Stability* are thus very closely related. These words describe in fact the same property of the clock generating the time scale.

There are some differences in the practical implementation between astronomical time scales and atomic time scales. The duration of the period of the phenomena used in astronomical timekeeping is very much longer than that of the oscillations used in laboratory clocks such as pendulum, crystal or atomic. In traditional astronomical timekeeping, clocks are therefore needed for subdivision of the observation periods (time interval between successive observations or groups of observations) into smaller and more practical intervals such as hours, minutes and seconds.

On the other hand, the very high frequency oscillations occurring in atomic clocks allow the generation of time scales by counting only. Furthermore, the various astronomical phenomena have periods which are different and not simply related:

Earth (rotation)	1 day
Moon	28 days
Earth (revolution)	1 year

These periods are not integer multiples of each other and this allows computations of configurations (epochs*) backwards into the past and forwards into the future. In astronomy, continuous counting of days is used throughout very long time intervals with an origin fixed by convention: the Julian Day numbers which are counted from an origin fixed on 1 January 4713 BC in the Julian Proleptic Calendar.

From the viewpoint of physical measurement of time, the operational differences between astronomical and physical (atomic) time scales do not seem to be fundamental. There has been, however, some misunderstanding and even some controversy between astronomers and physicists in the early years of atomic timekeeping, most of which being due to a conflict between old tradition and somewhat disrupting new ideas. In the old days, the concept of a "time scale" was not used. There was time and it had to be determined by observation of the stars. Clocks were neither accurate nor reliable and the idea of relying on clocks to define a time scale was thought to be risky and therefore rather unwise. Atomic clocks were accepted for defining better time intervals but not time. Not only in view of the rather poor reliability of early atomic frequency standards, but also from the basic experience of limited lifetime of all human works the question was a real one: What would happen if all clocks were destroyed? It is true that in such a case the time scale defined by these clocks would be lost forever. In practice, this problem has been overcome by redundancy and wide geographical distribution of timekeeping institutes. Astronomical observations have not been discontinued either. Therefore, the risk of losing time due to a cataclysm destroying all clocks is rather hypothetical, especially since such an event might as well leave nobody able to observe the starts.

At present, there is no longer any controversy as both types of time scales are continuously being kept and scientists working in position astronomy, celestial mechanics, geodesy, etc., would no longer dispute the convenience of accurate and reliable timing systems based on atomic clocks.

On the other hand, the angular position of the rotating Earth with reference to the stars continues to be required for the purposes of navigation. Several time scales are therefore required and short descriptions of the more important ones are given below:

---

* The word "epoch" as used in astronomy means a fundamental starting or reference time.[3] This meaning is derived directly from that of the ancient greek word ἐποχή which means "holding or reference point". In common language, epoch also means a long time interval. In some US literature on navigation, satellite tracking and geodesy, epoch has been used as a description of an instant on a time scale. This use is incorrect and alternatives have been suggested,[3,4] e.g. "date" and "clock time".

Examples of time scales

Universal Time
The time scales based on the rotation of the Earth are the following:

Sidereal Time. A sidereal day is the interval between two transits of the same star through the local meridian, i.e. the duration of one period of rotation with respect to the system of "fixed stars".

Mean Solar Time. "True Solar Time" as determined from the position of the Sun in the sky is not uniform throughout the year because of the motion of the Earth around the Sun. By averaging over one year, Mean Solar Time is obtained. The latter has a known relation to sidereal time: the ratio of a sidereal day to a mean solar day is roughly 1·00274.

Universal Time. Sidereal time at a given location on the Earth converted to mean solar time and referred to the meridian of Greenwich,* is called UT0, the uncorrected value of Universal Time. Applying corrections for the position of the pole result in UT1 which is the same on all points of the Earth, UT1 as a measure of the angular position of the Earth is important for navigation.
  Applying corrections for seasonal variations of the speed of rotation of the Earth results in UT2. However, there are unpredictable variations in the rotation of the Earth and therefore, over long intervals, UT2 is not significantly more uniform than UT1.

Ephemeris Time
ET is the uniform time scale used to determine the position of celestial bodies. The scale is defined by the orbital motion of the Earth about the Sun. The second of ET is defined as $1/31\ 556\ 923\ 9747$ of the tropical year for 0 January 1900. In practice, ET is obtained from the orbital motion of the Moon around the Earth. The computed position of the Moon with respect to the stars is tabulated as a function of ET in the Improved Lunar Ephemeris. An observed position of the Moon gives the ET of the epoch of observation of interpolation in the Improved Lunar Ephemeris. Since June 1952, the US Naval Observatory has been determining Ephemeris Time with the dual-rate Moon position camera.[5]
  At its XVIth General Assembly, Grenoble, 1976, the IAU adopted a new dynamical time scale for general astronomical use whose scale unit is the SI

---

  * The name Greenwich Mean Time GMT is still popular but should be replaced by the appropriate designations, either Universal Time (UTC) for general civil use or UT1 for celestial navigation or surveying.

second (see below) and which was aligned with the epoch of TAI on 1 January 1977. ET, as such, is now only of historical interest.

Atomic Time
Atomic Time (AT) is a time scale obtained by continuous counting of SI-seconds as defined since October 1967:

> "The second is the duration of 9 192 631 770 periods of the radiation corresponding to the transition between the two hyperfine levels of the ground state of the Cs^{133} atom".

By international agreement, the origin of atomic time scales has been set on 1 January 1958, on 0 h 0 m 0 s UT2 (BIH).

Since—as stated before—each clock defines its own time scale and no real clock is perfect, the time scales initially synchronized on the conventional origin will depart from each other after some time.

International coordination of timekeeping activities is needed to reach a common recognized time scale which is hoped to be more uniform than either of the individual time scales. The institution charged with the difficult task of establishing the *International Atomic Time Scale* (TAI) is the "Bureau International de l'Heure" (BIH), located at the Observatory of Paris, France. Some information on the international activities related to time scale generation and dissemination will be given in Section 4.3.

Coordinated Universal Time (UTC). Despite its name, UTC is a time scale generated by atomic clocks. The continuing requirement for a *time scale approximating UT* is due to its wide application in navigation. A compromise solution had therefore to be found which retains the advantage of uniform time scale generation by atomic clocks and still follows the variations of the rotation of the Earth.

The evolution of UTC has progressed in two phases. The first one was effective during the years 1961 to 1971 and was based on two corrective measures applied as needed and coordinated by the BIH:

(a) the basic frequency was offset, the offset remaining constant during at least one calendar year; and
(b) step adjustments of $\pm 0 \cdot 1$ s were introduced whenever needed to keep the difference UTC–UT2 as small as possible.

The frequency offsets were made with reference to the atomic frequency then already known but adopted only in 1967. Table 4.1 shows the values used.

The performance of this system, conceived to approximate UT2 as

TABLE 4.1

Year	Offset $\dfrac{f_{UTC} - f_{AT}}{f_{AT}} \times 10^{-10}$
1960	−150
1961	−150
1962	−130
1963	−130
1964	−150
1965	−150
1966	−300
1967	−300
1968	−300
1969	−300
1970	−300
1971	−300

Note: The negative sign means that the UTC-clock was slow compared to the AT-clock

closely as possible, was good, but the complexity of combined frequency offset and time step adjustments led to increased operational difficulties. These became especially apparent in sophisticated radionavigation systems as LORAN-C and OMEGA which are based on precise timing but have no relation to the rotation of the Earth. Frequency adjustments proved to be particularly awkward due to the large number of transmitters operating in navigation systems as well as those used for general time signal broadcasts. Therefore, after successful demonstration of systems operating on frequencies without offset and only time step adjustments and much discussion of the actual requirements of navigators using radio time signals for classical position determination, a revised scheme for UTC was adopted in the CCIR (see Section 4.3) and introduced on 1 January 1972. The new UTC definition has the following characteristics:

(a) There is *no offset* in the basic rate of UTC with respect to TAI.
(b) UTC will differ from TAI by an integer number of N seconds.
(c) The maximum allowed departure UT1–UTC is ± 0.9 seconds.*

For users requiring UT1 to a higher precision, time service system operating authorities may include a correction information in the time signal. The operational consequence of this definition is the necessity of introducing one second steps, at the end of the last day on a month. By analogy to the

* The maximum allowed departure UT1–UTC was ± 0.7 s in 1972. The tolerance has been increased to ± 0.9 s in 1974.[6] (See Appendix 8.2.)

intercalary day in a leap year in the calendar, this second step has been called "leap second", being regarded as positive when a second is added and negative when it is subtracted.

4.3. INTERNATIONAL COORDINATION

As already mentioned in the preceding section, international cooperation and coordination plays a dominant role in the field of standard time and frequency. Due to the inherent dynamic nature of time scale generation, these activities must go on without interruption. Furthermore, various aspects are dealt with by different organizations and bodies according to arising needs, competence, official status and tradition. References 1 and 2 give a detailed account on the evolution up to 1972. Since then, no major modification in the role of the international bodies has occurred. They form a complex system and the following short review cannot claim to reproduce accurately all details of their activity. The intent is only to give a general idea of their main purposes.

Sometimes, a distinction is made between governmental and non-governmental organizations. This is mainly formal, however, since all of them exist due to some kind of governmental support, with the exception that for the latter, the support is less direct.

Governmental organizations and agencies

CGPM   General Conference on Weights and Measures.
        International conference including the representatives of governments having signed the International Treaty on the Metre (1875, revised 1921).

CIPM   International Committee for Weights and Measures.
        Executive body of CGPM.

BIPM   International Bureau of Weights and Measures.
        (Address: Pavillon de Breteuil, F 92310-Sèvres, France.)
        Executive agency and laboratory operating under the authority of CIPM (CGPM).

CCDS   Consultative Committee for the Definition of the Second.
        Consultative body comprising scientists nominated by CIPM (created in 1956).

ITU   International Telecommunication Union.
        International union comprising the telecommunications' authorities and administrations of member countries.

CCIR   International Radio Consultative Committee.
            Consultative body of ITU dealing with radio communications. Its Study Group VII deals with standard frequency and time broadcasts.

Non-governmental scientific organizations
ICSU   International Council of Scientific Unions.

IAU   International Astronomical Union.
            Member of ICSU. IAU since its formation in 1919 was mainly responsible for all questions of time and still plays an eminent role through its Commission 31 (time).

IUGG   International Union of Geodesy and Geophysics.

URSI   International Radio Scientific Union.
            Deals with all questions of radio science. Its commission A (Electromagnetic Metrology) is involved in time and frequency.

FAGS   Federated Astronomical and Geophysical Services.
            A federal organization set up by the IAU and the IUGG to provide support for international observatory services in astronomy and geophysics that need to be maintained in continuous operation. Besides the BIH (below) the services of interest that it supports at this time are the International Latitude Service and the International Polar Motion Service.

BIH   Bureau International de l'Heure (International Time Bureau).
            International Agency operating as one of the services of FAGS under a Directing Board with representatives of the IAU, IUGG and URSI and observers appointed by CCIR and BIPM. (Address: 61 Av. de l'Observatoire, F 75014-Paris 14e, France.) The BIH is the international agency responsible for the TAI, UTC and UT time scales and publishes at regular intervals pertinent information on its operations.

On the side of the Governmental Organizations we have two main divisions, namely those responsible for the definition of basic units and methods: CGPM, CIPM, BIPM and the body responsible for ways and means of standard time and frequency dissemination with an optimum use of the radio frequency spectrum (Study Group 7 of the CCIR).

    On the other side we have URSI which has played a major role in some questions of fundamental nature, especially in recommending the revision of the UTC system in 1967 and last but not least, the BIH which, among the international bodies mentioned, is the only permanently active agency on the international level.

As shown in more detail in the references cited, the network of international organizations and agencies has proved to be very flexible and remarkably efficient in dealing with the considerable changes in the field of standard time and frequencies which have taken place during the past 20 years.

A recent and competent review of the basic concepts of precise time and frequency is contained in Ref. 7 which also includes the complete text of the most important international agreements and resolutions.

4.4. TIME SCALE GENERATION

Following the general ideas exposed in the preceding section we shall review here some practical aspects of time scale generation, timekeeping and time measurement. In common language, the word time has a double meaning, namely *time interval* (duration of a limited sequence of events) and time as a designation of an *instant on a time scale* ("clock time"). In experimental scientific work, time measurements are also used in both ways: sometimes, the information needed is only the duration of an event or the time interval between two events. In other cases, a greater number of events is to be recorded with reference to a time scale starting at the beginning of the experiment. In these two categories of experiments, knowledge of the starting time with respect to, for example, UTC is not needed with high precision. Finally, there is a third category of work where observations or measurements must be recorded or actions undertaken with reference to a common time scale such as UTC.

In this category, timing information in the form of a *display* or *time code* is required at the instants determined by the work programme. The required precision may vary within wide limits but what cannot be dispensed with is the continuity of the timing information which must always be available when needed.

Timing information can be obtained by two means: either by using an independent local clock or by using timing information transmitted from a remote clock. In most cases, both means are combined in some way.

Whatever its quality and expected performance, the independent clock cannot approximate the UTC time scale without being adjusted to the correct frequency and set to the correct time at the beginning of its operation. To do this, the clock must either be carried to a place where a reference time scale is available or remote timing information must be used, taking into account any possible time delay on the transmission path. The advantage of using a local clock is that the remote timing information is only occasionally needed for checks and readjustments, if necessary. Thus, the practical realization of a time scale by means of a local clock includes:

a reference time scale*;
frequency or rate adjustment; and
time adjustment or synchronization.

In the great majority of applications, it can be assumed that the errors of the reference time scale are negligible and we shall discuss this situation first. For the following discussion, it is insignificant whether the reference time scale is represented by a radio time signal obtained at the output of a receiver, a time code signal, a sequence of pulses or a visual display. The reading of a clock is the number of cycles of the basic clock frequency, counted from the beginning of the clock's operation. According to the definitions given in Section 2.1, the clocks reading $T(t)$ as a function of time t defined by the reference time scale is determined by the relation

$$T(t) = \frac{1}{v_0} \int_0^t v(t)\, dt, \qquad (4.1)$$

where v_0 is the reference clock frequency (assumed to be constant) and $v(t)$ the proper frequency of the local clock. Obviously, the reference clock reading $T_0(t)$ is

$$T_0(t) = t. \qquad (4.2)$$

Given a set of clocks to be compared to the reference clock, the reading $T_i(t)$ of the ith clock is

$$T_i(t) = t + \int_{t_{0i}}^t y_i(t)\, dt - t_{si} \qquad (4.3)$$

where y_i is the normalized frequency offset as defined by Eq. (2.5) and t_{si} the delay possibly introduced at the synchronization of the ith clock, initiated at time t_{0i} as read on the reference time scale. The difference between the readings of the ith clock and the reference is thus equal to:

$$x_i(t) = T_i(t) - T_0(t) = \int_{t_{0i}}^t y_i(t)\, dt - t_{si}. \qquad (4.4)$$

___

* Here we do not discuss the exceptional case of a highly accurate atomic clock taking part in the formation of the TAI and UTC scales. In that application, the TAI or UTC scales are also reference scales but the clock must not be adjusted with respect to that reference.

This equation is very similar to Eq. (2.6), the only differences are the introduction of the synchronization delay t_{si} and the lower limit t_{0i} instead of zero in the integral, introduced because we start operating at time t_{0i} and not at time zero.

The synchronization delay t_{si} can be made very small by subsequent comparisons and time adjustments so that in most applications, it will remain small and will soon be dominated by the first term.

In order to avoid confusion in recording results of clock comparisons, it is very important to state clearly which is the reference and which the clock to be compared. This is the only way to exchange meaningful results between different workers in the field. Notation and sign conventions have been agreed upon internationally through CCIR Recommendation 459 (1970), the text of which is reproduced in Appendix 1 to this chapter. The equations and definitions adopted in this section are consistent with this agreement.

By observing the behaviour of a real clock as described by Eq. (4.4), it is customary to distinguish between *deterministic* and *random* effects upon $x_i(t)$. This has already been mentioned in Section 2.2 and it has become customary to treat as deterministic phenomena in clock operation the two parameters of Eq. (2.8), namely initial offset and linear drift. If we transform the variables to comply with the initial conditions defined in (4.3), we obtain for the normalized frequency

$$y_i(t) = y_{ri} + a(t - t_{0i}) + y_{0i} \qquad (4.5)$$

where a is the normalized drift rate and y_{0i} the initial offset. All remaining random or unspecified fluctuations are represented by y_{ri} as discussed in Chapter 2. Inserted in (4.4), we obtain for the time difference:

$$x_i(t) = \int_{t_{0i}}^{t} y_{ri}(t)\,dt + \frac{a}{2}(t - t_{0i})^2 + y_{0i}(t - t_{0i}) - t_{si}. \qquad (4.6)$$

The reason for distinguishing between deterministic and random effects in the way shown here is due more to practical than fundamental reasons. One could imagine some more complex effects than just linear drift to be deterministic in principle, e.g. cyclic temperature variations. However, such effects are observed only in special cases, and in current practice it is preferred to have the general model as simple as possible.

The main problem of observing, analysing and characterising the behaviour of a clock resides in the fact that $x_i(t)$ on the left-hand side of (4.6) is the most easily measured quantity. The task of the experimenter is to try to determine the parameters of the right-hand side from repeated measurements of x_i. In many cases, this is very simple but there are examples where it becomes difficult to distinguish between small frequency jumps, changing drift or low-frequency divergent noise. We shall illustrate this by the examples given below. Furthermore, the need for clock modelling or characterization becomes evident from the users' requirements which are usually stated in the form of a maximum allowable timing error.

Example 1

As the first of two examples, we shall discuss the performance of a good-quality quartz crystal oscillator. Except for very short time intervals in which we are not interested here, the effects of random noise do not appear in the results. Temperature influence can be reduced to insignificance by a good temperature controlled oven. It is then possible to adjust the initial frequency offset within very small limits (parts in 10^{11}) so that the ageing remains as the dominant parameter for the description of the oscillator's performance as a clock.

We assume that the frequency of the oscillator is adjusted to $1 \cdot 0$ MHz and that the oscillator has been kept running for some time before, so that the initially rapid ageing has settled down to a constant linear frequency drift of 1×10^{-9} per day. The oscillator drives a frequency divider producing an output of 1 pulse per second. We determine the time difference between this pulse and our laboratory clock pulse by means of a time interval counter (see Chapter 5).

We set $t = 0$ at the beginning of the experiment and assume negligible synchronization error. Then, the time difference $x_1(t) = T_1 - T_0$ is given by

$$x_1(t) = \frac{a}{2} t^2 \text{ (s)} \tag{4.7}$$

with $a = \dfrac{10^{-9}}{86\,400} \text{ s}^{-1}$ and t expressed in seconds.

Table 4.2 shows the results for the first ten days with t expressed in days, 1 day = 86 400 s. The third column shows the first differences of the time readings.

$$\Delta_1 x_1 = x_1(t + \tau) - x_1(t)$$

with $\tau = 1\,\text{d} = 86\,400$ s. From these values, we can obtain the average frequency offsets over the measurement intervals

$$\bar{y}_1(t, \tau) = \frac{\Delta_1 x_1}{\tau}$$

e.g. after the first day $\bar{y}_1(1, \tau) = +0 \cdot 5 \times 10^{-9}$, whereas the (instantaneous) offset after one day is $y_1(1) = +1 \times 10^{-9}$ (see last column).

TABLE 4.2

t (days)	x_1 (μs)	$\Delta_1 x_1$ (μs)	$\Delta_2 x_1$ (μs)	y_1 (1×10^{-9})
0	0			0
		43·2		
1	43·2		86·4	1
		129·6		
2	172·8		86·4	2
		216·0		
3	388·8		86·4	3
		302·4		
4	691·2		86·4	4
		388·8		
5	1080·0		86·4	5
		475·2		
6	1555·2		86·4	6
		561·6		
7	2116·8		86·4	7
		648·0		
8	2764·8		86·4	8
		734·4		
9	3499·2		86·4	9
		820·8		
10	4320·0			10

The fourth column shows the second differences

$$\Delta_2 x_1 = x_1(t + 2\tau) - 2x_1(t + \tau) + x_1(t)$$

which are directly related to the drift rate

$$a = \frac{\Delta_2 x_1}{\tau} = \frac{86·4\ \mu s}{86\,400} = 10^{-9}.$$

In this example, the time difference has accumulated to $+4·32$ ms after 10 days and the frequency offset is $+10^{-8}$.

If we now make a downward frequency adjustment by -2×10^{-8} but no time adjustment, the time difference $x_1(t)$ will evolve as follows:

Let $t_1 = t - 10$ days and formulate a new equation for x_1:

Using (4.6), we have $y_{01} = -1 \times 10^{-8}$ due to the readjustment and $t_{s1} = -4·32$ ms and thus

$$x_1(t_1) = \frac{a}{2}t_1^2 - 10^{-8}t_1 + 4·32 \times 10^{-3} \quad \text{(s)} \qquad (4.8)$$

TABLE 4.3

t (days)	t_1 (days)	x_1 (μs)	$\Delta_1 x_1$ (μs)	$\Delta_2 x_1$ (μs)	y_1 (1×10^{-9})
10	0	4320·0			-10
			820·8		
11	1	3499·2		86·4	-9
			734·4		
12	2	2764·8		86·4	-8
			648·0		
13	3	2116·8		86·4	-7
			561·6		
14	4	1555·2		86·4	-6
			475·2		
15	5	1080·0		86·4	-5
			388·8		
16	6	691·2		86·4	-4
			302·4		
17	7	388·8		86·4	-3
			216·0		
18	8	172·8		86·4	-2
			129·6		
19	9	43·2		86·4	-1
			43·2		
20	10	0			0

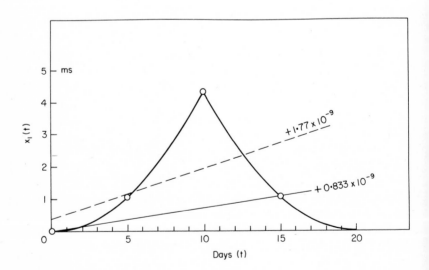

Fig. 4.1.

Table 4.3 shows the daily readings for the next ten days, at the end of which x_1 and y_1 have come back to zero. The diagram in Fig. 4.1 shows the results in graphical form.

It is obvious that a real oscillator will not behave exactly in this way because of the impossibility of setting frequency and time exactly. The purpose of this artificial example is to illustrate some elementary computational techniques, especially the use of first and second differences of time difference readings for estimating the average frequency offset and the drift rate respectively.

The use of finite differences of phase or phase-time measurements has been suggested by J. A. Barnes[8] as a simple but powerful technique for the analysis of clock errors, especially random fluctuations with low-frequency divergent spectral densities such as flicker and random walk.

In our simple deterministic example, these computations seem to be almost trivial, but we can use the example to draw attention to a possible pitfall in the analysis of noisy results with the purpose of estimating the average frequency offset over a given measurement time interval:

By definition (2.6), the phase-time $x(t)$ is the time integral of the normalized frequency offset $y(t)$. The time average of any reasonably well-behaved function $y(t)$ in the interval $t_1 \leqslant t \leqslant t_2$ is

$$\overline{(y(t))}_{t_1, t_2} = \frac{1}{t_2 - t_1} \int_{t_1}^{t_2} y(t) \, dt \qquad (4.9)$$

and thus

$$\overline{(y(t))}_{t_1, t_2} = \frac{x(t_2) - x(t_1)}{t_2 - t_1}. \qquad (4.10)$$

Let us take the results in Tables 4.2 and 4.3 and Fig. 4.1 and assume that we have readings only for $t = 0, 5, 10, 15$ days. We record these points graphically in Fig. 4.2. The average frequency offset over the whole 15-day period is obviously:

$$(y_1)_{0 \cdot 15} = \frac{1 \cdot 08 \times 10^{-3} - 0}{15 \times 86\,400} = 0 \cdot 8333 \times 10^{-10}$$

as shown by the slope of the solid straight line. Assume now we have only these four points and no further knowledge about the way in which they were generated. Then, the data look like being random and as well-educated experimenters, we try to fit a straight line by means of the well-known

method of least squares. This leads to the dotted straight line described by
the equation

$$x_a(t) = (0.1525\,t + 0.3812) \times 10^{-3}\ \text{s}$$

with t in days. The slope corresponds to a "least square fit"-frequency offset
of $+ 1.77 \times 10^{-9}$. Should we use this as an estimate for the average frequency
offset, we would be in error by 112%.

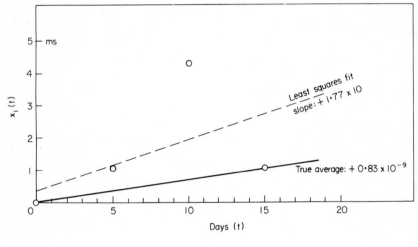

F$_{\text{IG}}$. 4.2.

The computational effort needed to obtain this incorrect result is roughly
ten times greater than that yielding the correct answer. This example has of
course been chosen to show a large error. A similar comparison made over
days 0 to 20 or 0 to 10 produces accidentally a correct result. Incidentally,
the large error observed is consistent with the theory of least square fitting.
If we compute the standard deviation of the slope, based on the four data
points, we find a value of 0.14, i.e. the slope falls within the interval 0.1525 ±
0.14. This example shows a case of evident misuse of least square fitting. It
has been chosen with intention to be somewhat exaggerated. In real measurements on oscillators and clocks including noise, the errors are less obvious.
The method of least squares fails in the example because of the strong
correlation between the successive readings of the phase which are strongly
correlated through a simple law. There is no random process present in this
case and the number of samples is so small that the statistical significance is
poor.

TABLE 4.4

Date (1500 UTC)	MJD	x (μs)	$\Delta_1 x$ (μs)	$\Delta_2 x$ (μs)	y ($\times 10^{-12}$)
7. 1.1973	41 689	29·2			
			−0·9		−1·04
17. 1	699	28·3		+0·3	
			−0·6		−0·69
27. 1	709	27·7		−0·4	
			−1·0		−1·15
6. 2	719	26·7		+0·1	
			−0·9		−1·04
16. 2	729	25·8		−0·5	
			−1·4		−1·62
26. 2	739	24·4		+0·1	
			−1·3		−1·50
8. 3	749	23·1		+0·3	
			−1·0		−1·15
18. 3	759	22·1		+0·5	
			−0·5		−0·58
28. 3	769	21·6		−0·9	
			−1·4		−1·62
7. 4	779	20·2		−0·2	
			−1·6		−1·85
17. 4	789	18·6		−0·1	
			−1·7		−1·97
27. 4	799	16·9		+0·6	
			−1·1		−1·27
7. 5	809	15·8		+0·3	
			−0·8		−0·93
17. 5	819	15·0		−2·2	
			−3·0		−3·47
27. 5	829	12·0		+2·8	
			−0·2		−0·23
6. 6	839	11·8		−2·2	
			−2·4		−2·78
16. 6	849	9·4		+2·1	
			−0·3		−0·35
26. 6	859	9·1		+0·1	
			−0·2		−0·23
6. 7	869	9·3		−0·3	
			−0·5		−0·58
16. 7	879	8·8		−3·1	
			−3·6		−4·16
26. 7	889	5·2		+0·4	
			−3·2		−3·70
5. 8	899	2·0		+3·7	
			+0·5		+0·58
15. 8	909	2·5		+1·7	
			+2·2		+2·55

E

TABLE 4.4—*continued*

Date (1500 UTC)	MJD	x (μs)	$\Delta_1 x$ (μs)	$\Delta_2 x$ (μs)	y ($\times 10^{-12}$)
25. 8. 1973	41 919	4·7		$-0·1$	
			$+2·1$		$+2·43$
4. 9	929	6·8		$-0·1$	
			$+2·0$		$+2·31$
14. 9	939	8·8		$-0·4$	
			$+1·6$		1·85
24. 9	949	10·4		$+0·3$	
			$+1·9$		2·20
4.10	959	12·3		0	
			$+1·9$		2·20
14.10	969	14·2		0	
			$+1·9$		2·20
24.10	979	16·1		$+0·3$	
			$+2·2$		2·55
3.11	989	18·3		$-0·2$	
			$+2·0$		2·31
13.11	999	20·3		$-0·1$	
			$+2·1$		2·43
23.11	42 009	22·4		$-0·5$	
			$+1·6$		1·85
3.12	019	24·0		$-0·5$	
			$+1·1$		1·27
13.12	029	25·1		$-0·2$	
			$+0·9$		1·04
23.12	039	25·8		$-0·3$	
			$+0·6$		0·69
2. 1. 1974	049	26.4		$+0·5$	
			$+1·1$		1·27
12. 1	059	27·5		$+0·2$	
			$+1·3$		1·50
22. 1	069	28·8		$+0·3$	
			$+1·6$		1·85
1. 2	079	30·4		$-2·3$	
			$-0·7$		$-0·81$
11. 2	089	29·7		$+1·7$	
			$+1·0$		1·15
21. 2	099	30·7		$+0·7$	
			$+1·7$		1·97
3. 3	109	32·4		$-0·3$	
			$+1·4$		1·62
13. 3	119	33·8		$-0·6$	
			$+0·8$		0·93
23. 3	129	34·6		$-0·5$	
			$+0·3$		0·35

TABLE 4.4—*continued*

Date (1500 UTC)	MJD	x (μs)	$\Delta_1 x$ (μs)	$\Delta_2 x$ (μs)	y ($\times 10^{-12}$)
2. 4. 1974	42 139	34·9		+0·2	
			+0·5		0·58
12. 4	149	35·4		+0·4	
			+0·9		1·04
22. 4	159	36·3		−1·5	
			−0·6		−0·69
2. 5	169	35·7		−0·1	
			−0·7		−0·81
12. 5	179	35·0		+0·4	
			−0·3		−0·35
22. 5	189	34·7		−0·7	
			−1·1		−1·27
1. 6	199	33·6		−1·8	
			+0·7		+0·81
11. 6.	209	34·3		0	
			+0·7		+0·81
21. 6	219	35·0		−0·5	
			+0·2		0·23
1. 7	229	35·2		+0·7	
			+0·9		1·04
11. 7	239	36·1		+0·5	
			+1·4		1·62
21. 7	249	37·5		+0·1	
			+1·5		1·74
31. 7	259	39·0		+0·8	
			+2·3		2·66
10. 8	269	41·3		−1·0	
			+1·3		1·50
20. 8	279	42·6		−0·2	
			+1·1		1·27
30. 8	289	43·7		−1·1	
			0		0
9. 9	299	43·7		0	
					0
19. 9	309	43·7	0	−1·1	
			−1·1		−1·27
29. 9	319	42·6		0	
			−1·1		−1·27
9.10	329	41·5		−0·1	
			−1·2		−1·39
19.10	339	40·3		+0·1	
			−1·1		−1·27
29.10	349	39·2		−0·3	
			−1·4		−1·62

TABLE 4.4—*continued*

Day (1500 UTC)	MJD	x (µs)	$\Delta_1 x$ (µs)	$\Delta_2 x$ (µs)	y ($\times 10^{-12}$)
8.11.1974	42 359	37·8		+0·1	
			−1·3		−1·50
18.11	369	36·5		−0·1	
			−1·4		−1·62
28.11	379	35·1		+0·1	
			−1·3		−1·50
8.12	389	33·8		−0·2	
			−1·5		−1·74
18.12	399	32·3		−0·5	
			−2·0		−2·31
28.12	409	30·3		+0·2	
			−1·8		−2·08
7. 1.1975	419	28·5		−0·1	
			−1·9		−2·20
17. 1	429	26·6		−0·4	
			−2·3		−2·66
27. 1	439	24·3		0	
			−2·3		−2·66
6. 2	449	22·0		−0·2	
			−2·5		−2·89
16. 2	459	19·5		+2·0	
			−0·5		−0·58
26. 2	469	19·0		+4·1	
			+3·4		+3·94
8. 3.	479	22·4		−3·6	
			−0·2		−0·28
18. 3	489	22·2		−0·6	
			−0·8		−0·93
28. 3	499	21·4		−0·5	
			−1·3		−1·50
7. 4	509	20·1		−0·2	
			−1·5		−1·74
17. 4	519	18·6		+0·5	
			−1·0		−1·16
27. 4	529	17·6		+1·5	
			+0·5		+0·58
7. 5	539	18·1		+0·2	
			+0·7		0·81
17. 5	549	18·8		−0·1	
			+0·6		0·69
27. 5	559	19·4		+0·3	
			+0·9		1·04
6. 6	569	20·3		−1·8	
			−0·8		−0·93

TABLE 4.4—*continued*

Day (1500 UTC)	MJD	x (μs)	$\Delta_1 x$ (μs)	$\Delta_2 x$ (μs)	y ($\times 10^{-12}$)
16. 6. 1975	42 579	19·5		−0·6	
			−1·4		−1·62
26. 6	589	18·1		−0·3	
			−1·7		−1·97
6. 7	599	16·4		+0·3	
			−1·4		−1·62
16. 7	609	15·0		0	
			−1·4		−1·62
26. 7	619	13·6			

Example 2

In this example, the long-term performance of commercial caesium frequency standards is discussed. As the reference time scale we use a LORAN-C signal received by means of a special LORAN-C timing receiver. Phase-time readings are taken daily at 1500 UTC. The characteristics of the LORAN-C systems are described in Chapter 8 and in the literature cited there. The transmitter used in this example is the W-slave station of the Norwegian Sea chain, located on the island of Sylt, off the coast of the northern part of Germany and the receiver is located in Bern, Switzerland, about 800 km south of the transmitter. The transmitter frequency and pulse rate are controlled by a caesium standard whose performance is known with reference to the US Naval Observatory time scale by means of published daily phase corrections. These corrections are not used in this example since the phase-time variations of the LORAN-C chain are much smaller than those of the local caesium standard in Bern. Short-term phase fluctuations due to propagation delay variations and to noise in the receiver lead to an uncertainty of the daily readings of not more than ± 100 ns. Here we are not interested in these short-term variations and, for the discussion of the slow variations due to the caesium standard, we need only readings every ten days. These are given in Table 4.4.

In the first column we have the calendar date. The second column gives the MJD number (Modified Julian Day), a continuous day count which is more practical for computation and graphical representation. The dates chosen are those used by BIH for international time comparison. The third column gives the phase-time difference $x = \text{PTCH*−SYLT}$ in microseconds, according to the CCIR sign convention (see Appendix 1 to this chapter), i.e. a positive value of x means that the clock reading of PTCH is higher than that of SYLT.

* PTCH is the standard at Bern.

The fourth and fifth columns give the first and second differences of the x-values and the sixth column the average normalized frequency difference for the period between the two corresponding dates defined (see Chapter 2) as follows

$$\bar{y}_k(t_k, \tau) = \frac{1}{\tau}(x(t_k + \tau) - x(t_k)) = \frac{\Delta_1 x}{\tau} \tag{4.11}$$

For the period between MJD 41 689 and 41 699, we have thus:

$$y = \frac{-0.9 \times 10^6}{10 \times 8.64 \times 10^5} = -1.04 \times 10^{-12}.$$

The second differences are used to calculate the Allan-variance. It is easily shown that

$$\bar{y}_{k+1} - \bar{y}_k = \frac{1}{\tau}(x(t_k + 2\tau) - 2x(t_k + \tau)) + x(t_k)) = \frac{\Delta_2 x}{\tau} \tag{4.12}$$

and thus, according to the definition of Eq. (2.27), we have

$$\sigma_y^2(\tau) = \frac{1}{2\tau^2}\langle(\Delta_2 x)^2\rangle. \tag{4.13}$$

Squaring and adding all $M = 92$ values of $\Delta_2 x$ in Table 4.4, we have

$$\langle(\Delta_2 x)^2\rangle = \frac{115.19 \times 10^{-12}}{92} = 1.252 \times 10^{-12} \, \text{s}^2$$

and

$$\sigma_y^2 = \frac{1.252 \times 10^{-12}}{2 \times (8.64 \times 10^5)^2} = 8.386 \times 10^{-25}$$

and

$$\sigma_y(\tau = 10 \, \text{days}) = 9.16 \times 10^{-13}.$$

Similarly, we may compute $\sigma_y(\tau)$ for larger values of τ. This has been done and the results are given in Table 4.5.

TABLE 4.5

$\tau =$	1	10	20	50	100 days
$\sigma_y(\tau) =$	9.45	9.16	8.72	8.2	9.97×10^{-13}

TABLE 4.6

MJD	x (μs)	$\Delta_1 x$	$\Delta_2 x$
\multicolumn{4}{c}{(November 1974)}			
42 352	38·7		
		−0·1	
53	38·6		
		−0·1	0
54	38·5		
		−0·1	−0·1
55	38·4		
		−0·2	+0·1
56	38·2		
		−0·1	0
57	38·1		
		−0·1	−0·1
58	38·0		
		−0·2	+0·1
59	37·8		
		−0·1	0
60	37·7		
		−0·1	0
61	37·6		
		−0·1	−0·2
62	37·5		
		−0·3	+0·3
63	37·2		
		0	−0·1
64	37·2		
		−0·1	−0·1
65	37·1		
		−0·2	+0·1
66	36·9		
		−0·1	−0·1
67	36·8		
		−0·2	+0·1
68	36·6		
		−0·1	0
69	36·5		
		−0·1	−0·1
70	36·4		
		−0·2	+0·1
71	36·2		
		−0·1	0
72	36·1		
		−0·1	−0·1
73	36·0		
		−0·2	0
74	35·8		
		−0·2	

TABLE 4.6—*continued*

MJD	x (μs)	$\Delta_1 x$	$\Delta_2 x$
423 75	35·6		+ 0·2
		0	
76	35·6		− 0·2
		− 0·2	
77	35·4		+ 0·1
		− 0·1	
78	35·3		− 0·1
		− 0·2	
79	35·1		+ 0·1
		0	
80	35·1		− 0·1
		− 0·1	
81	35·0		

The value for $\tau = 1$ day has been computed from data which are not given in Table 4.4, but Table 4.6, i.e. 27 samples from November 1974.

The phase-time difference $x(t)$ is also represented graphically in Fig. 4.3 in order to give a general idea about the behaviour of the clock. Figure 4.4 shows graphically the data of Table 4.5.

The error bars show the standard deviation of the variance $\sigma_y^2(\tau)$, calculated from the data samples of Tables 4.4 and 4.6. It appears from this figure that over the range of 1 day $< \tau < 100$ days, $\sigma_y(\tau)$ remains constant within the error limits, at an average value of $9\cdot1 \times 10^{-13}$.

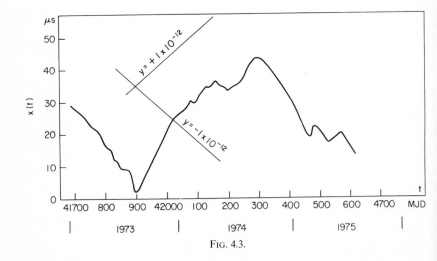

FIG. 4.3.

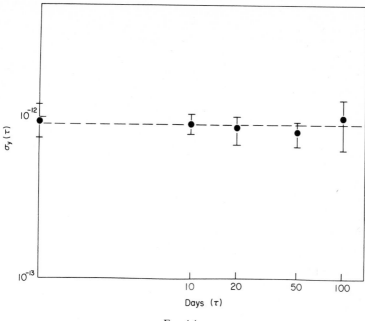

FIG. 4.4.

As discussed in Chapter 2, this type of fluctuation is described as flicker ($1/f$)-noise frequency modulation. The coefficient K_α of Eq. (2.35) is $K_\alpha = (9\cdot1 \times 10^{-13})^2 = 8\cdot28 \times 10^{-25}$ and, according to Table 2.1, the spectral density for very low Fourier frequencies is described by:

$$S_y(f) = \frac{8\cdot28 \times 10^{-25}}{2 \ln 2 \times f} = \frac{5\cdot97}{f} \times 10^{-25} \text{ s}$$

for $f < \sim 10^{-5}$ Hz.

Looking at Fig. 4.3, one is tempted to believe that the clock has made two major frequency jumps (around MJD 41 900 and 42 300) and only minor variations in between. Some deterministic cause for these jumps should have been found. This has not been possible on this particular clock operated since late 1972 in the author's laboratory. What we know from results on the many caesium clocks available from the BIH is that this clock shows larger variations than the best clocks in the TAI System but remains well within the $\pm 1 \times 10^{-11}$ limit for long-term stability claimed by its manufacturers.

The purpose of this example however, is not to present results of a particularly well-kept high-performance clock but to show what can be expected from an average apparatus operated without special precautions and checked by means of a LORAN-C receiver, thereby illustrating the practice of numerical computations involved with this type of operation.

After having established the presence of a flicker noise behaviour as shown in Fig. 4.4, further attempts to seek deterministic causes for the apparent frequency "jumps" have been abandoned. It is one of the various aspects of the elusive character of flicker noise phenomena that such causes cannot be found. This has been illustrated by the recent work of D. W. Allan[9] and H. Brandenberger,[10] i.e. that computer simulations of integrated flicker noise variables produce time functions having close similarity to that of Fig. 4.3. The basic question of the physical origin of the $1/f$-noise phenomena remains still open. However, the level of the horizontal floor in the $\sigma_y(\tau)$ representation is a good measure for the long-term stability of a high-performance clock.

It is worth noting that on the clock used in this example, no initial frequency adjustment with respect to an external reference had been made when it was put into service in November 1972. One readjustment of -2×10^{-12} was made on MJD 42481 on account of a problem with the C-field supply, a problem which had been noticed a few days before and which was corrected without interrupting the service. It was possible to have the clock free running since there was no application requiring an accuracy in frequency better than $\pm 1 \times 10^{-11}$.

The actual average frequency deviations can be roughly estimated by comparison to the indicated slopes corresponding to $y = \pm 1 \times 10^{-12}$ in Fig. 4.3. As mentioned before, average frequency differences for any given period of time are easily computed from the phase time differences accumulated during that period, whereas least square fitting of a straight line should not be used (see Example 1).

Figure 4.3 gives an example of timekeeping performance which can be expected from a free running caesium clock operating in a good but not especially controlled laboratory environment (no temperature control, slow variations within $\pm 5°C$ throughout the year).

Higher performance, i.e. a more uniform time scale can be obtained by two means, namely:

(a) Operating several independent clocks and constructing an average time scale.
(b) Periodic calibration and correction with reference to a high-performance laboratory standard.

In the official time services both methods are applied. The calibration and correction method was very popular in the early years of atomic timekeeping (1955–1966) when atomic standards were not reliable enough to be operated continuously as clocks. Crystal oscillators were used at that time for the physical generation of the timing signals and the error due to the ageing was corrected by means of periodic frequency calibrations against an atomic frequency standard. Nowadays, there are only a few high accuracy laboratory standards which are significantly more accurate and stable than commercial instruments. These are not operated continuously, not because of their lack of reliability but on account of periodic re-evaluation of the error budget determining the operational accuracy. Periodic calibration of an ensemble of clocks keeping the laboratory time scale is again used in this case. It is worth mentioning, however, that the primary standard at the National Research Council, Ottaway, Canada, has been operated continuously as a clock since May 1975.

The results of the calibrations are used to compute time corrections to the indicated clock times. The clocks are usually not readjusted physically except for initial synchronization and frequency setting of a new clock brought into the ensemble. Time scales built up in this way are often called "paper time scales" since the timing information is available only on paper after having added the computed correction to the result of the physical measurement.

The main difference between the methods (a) and (b) is that (a) allows the generation of a more uniform and reliable time scale than a single clock, i.e. the *stability* of the average frequency is better but the *accuracy* cannot be much better than that of a single clock.

Calibration with reference to a more accurate standard (method (b)) improves both the uniformity and the accuracy of the time scale.

The individual members of a set of clocks do not necessarily have equal performance. One bad performer in the set might offset the purpose of the averaging operation and it has therefore become customary to compute a *weighted average* of the individual time scale. The weights are determined from past performance determined by means of comparisons within the set, used for a limited prediction interval and periodically revised. The set of rules to establish the weights, the evaluation and prediction intervals is called a "time scale algorithm". There are many possible ways of establishing such algorithms and research work in this field is still going on as the state of the art in clock design and operation continues to improve.

The simplest method of attributing weights to clocks in a set is to assign a weight 0 or 1 to each clock, i.e. one uses only a subset of equal weight, discarding the remainder.[11] The problem, of course, is then to have a criterion to decide whether 0 or 1 should be applied. In Ref. 11, a very pragmatic approach is used, giving weight one to the 10 to 16 best performers of a set of over 30 clocks.

An example of a time scale algorithm using unequal weighting is the ALGOS method developed at the BIH and described in the 1973 annual report of this organization.[12] In this method, each weighting coefficient is proportional to the inverse of a variance* of the clock's rate measured over a period of 2 months. The whole set includes about 80 clocks located at various laboratories and these are compared via LORAN-C signal receptions. In order to avoid that only two or three clocks, which, by accident, showed exceptional performance during a 2-month interval dominate the system, the maximum (normalized) weight coefficient is limited to 100, the normalization constant being chosen so that the average weight of the clocks in the set is about 50.

We shall not go into more detail on time scale algorithms here since research on this subject is being pursued.

The optimum calibrations and corrections according to method (b) are also still under investigation. The error of an individual calibration depends on the statistical properties of the clock to be calibrated. These properties also determine the error of the clock time prediction; therefore, the development of reliable statistical clock models is important for the generation of accurate and uniform time scales.

If the parameters of the clock oscillator are known, e.g. the values of h_α in Eq. (2.33) for the spectral density $S_y(f)$, the problem is to find the optimum operational rules, i.e. how often a calibration is needed and how long each calibration should last. This problem has been investigated by several authors.[13, 14] The problem of detecting "abnormal" changes in clock performance, i.e. frequency changes and drift which are larger than the values to be expected from the known flicker noise parameter h_{-1} has been discussed in Ref. 15. It is shown there that even elaborate statistical tests cannot guarantee a high reliability in detecting frequency changes and jumps in the order of magnitude of the $\sigma_y(\tau)$ in the flicker noise region. This appears to confirm the remarks made above in the discussion of Example 2.

4.5. TIME CODES

The conventional visual time displays, either in analogue form as the traditional clock dial with hour, minute and second hands or the more modern digital displays are by themselves useful only for limited precision work involving human operators.

Errors of a few tenths of a second are inevitable on account of the variations of operator reaction time. Thus, for all measurements requiring higher

---

* The variance used is σ_y ($N = 6$, $T = \tau = 2$ months).

precision of timing, the influence of human operators must be eliminated. This is best done by using electrical signals. A fast logic level transition can provide timing resolution of a few nanoseconds or better. Time interval measurements can be made with corresponding resolution using fast counters (see Chapter 5). Complete time information, e.g. with reference to UTC, i.e. precise dating of an event, requires some means of transferring the clock display information to the permanent record kept for the experiment.

There are many possible ways of doing this transfer of timing information in the form of a set of rules defining this information. A time code is such a set of rules. As an example, let us have a digital display giving

(a) the number of the day in the year	3 digits
(b) the hours	2 digits
(c) the minutes	2 digits
(d) the seconds	2 digits
	9 digits

and changing state at the beginning of each second, it is possible to produce the information as binary coded decimal logic levels for each digit, i.e. a total of $4 \times 9 = 36$ bits on 36 lines simultaneously. This is an example of a *parallel code*, useful for applications involving computer systems. The code is available for transfer or recording into the connected data system during each second following the change of state and its resolution is one second. Higher resolution can be obtained by starting an additional counter at the beginning of each second. The event to be recorded must then do the following actions:

(1) Stop the counter.
(2) Record the time on the display.
(3) Record the counter display.
(4) Reset the counter to zero.

The next second's pulse starts the counter again. This setup is useful for recording events which do not occur more frequently than once every second. The resolution is limited by the counter time base frequency, e.g. to ± 100 ns with a 10 MHz counter. The recorded time of the event is the sum of the numbers on the time display and the numbers on the counter display. Problems could arise if the event occurs just at the time of change of state, i.e. at one full second, but they can be avoided by letting the counter overflow and by proper organization of the switching system used for the transfer of the displayed data.

A major disadvantage of parallel codes is the requirement of one line or connection per bit. Dissemination, transmission and recording of timing information in parallel form is rather uneconomical as shown by the above given example.

Timing information to be used at various points in extended systems is therefore best transmitted sequentially in the form of a *serial time code*.

Again, a tremendous variety of time code structures can be and has been developed. We shall therefore limit our discussion to a few principles and general ideas.

Among serial time codes, a distinction can be made between rudimentary and complete codes. A rudimentary code conveys only partial information which is completed by other means. A most simple form is a sequence of seconds pulses with an additional minute marker. This code is very popular among radio time signal services and very useful for many applications.

Complete time information up to one minute can, in most cases, be easily obtained by simple clocks and even good modern watches and waiting for the next minute marker is acceptable to all but the most impatient experimenters.

The more elaborate complete serial time codes have been developed mainly for use with automatic data acquisition and processing equipment.

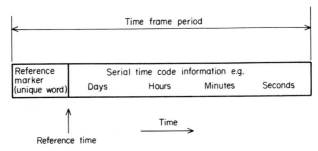

FIG. 4.5. Basic structure of a serial time code.

Figure 4.5 shows the basic structure of a serial time code. It is composed of a *reference marker* which must have a special structure *not* possible in the following information code, so that the logic transition marking the *reference time* can be unambiguously detected by appropriate logic circuitry. After the reference marker follows the sequence of coded pulses which contain the timing information. The duration of the whole sequence, including the reference marker, is called the *time frame*. The choice of the time frame period and its inverse, the time frame repetition rate, depends on the application. In standard time code formats, time frame periods between 0·1 s and 1 hour are used most frequently. Some effort of standardization has been made but in 1965 already, over 30 different time code formats were in use in the United States on missile ranges and similar installations.

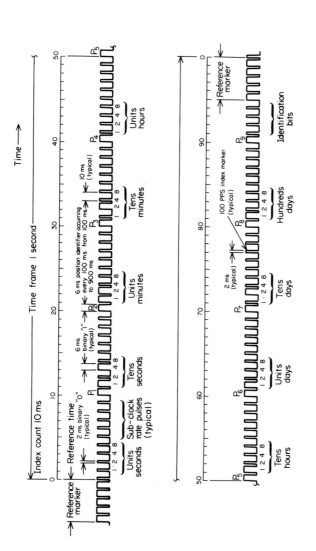

FIG. 4.6. NASA 36 Bit time code 100 p.p.s.

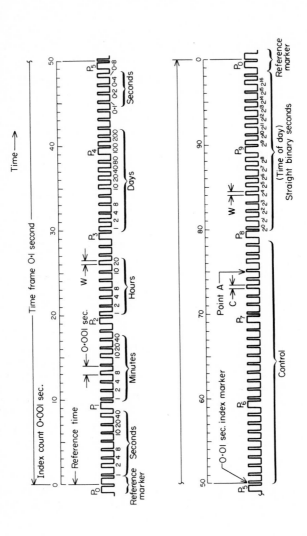

FIG. 4.7. IRIG standard format A 1000 p.p.s.

P = Position identifier, 0·8 ms duration
W = Weighted code digit, 0·5 ms duration
C = Control element (example) 0·5 ms duration
Duration of index markers = 0·2 ms

Time at point A: 21:18:42 + 0·8 + 0·07 + 0·005
= 21 hr, 18 min, 42·857 sec. on day 173

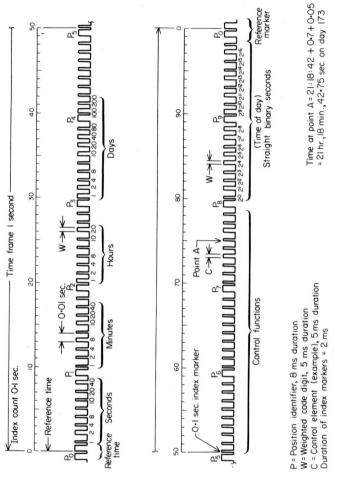

FIG. 4.8. IRIG standard format B 100 p.p.s.

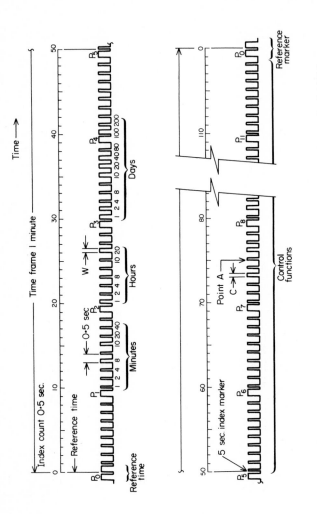

P = Position identifier, 0·4 sec. duration
W = Weighted code digit, 0·25 sec. duration
C = Control element (example) 0·25 sec. duration
Duration of index markers = 0·1 sec.

Time at point A = 21:18 + 37·5 sec.
= 21 hr, 37·5 sec. on day 173

FIG. 4.9. IRIG standard format C 2 p.p.s.

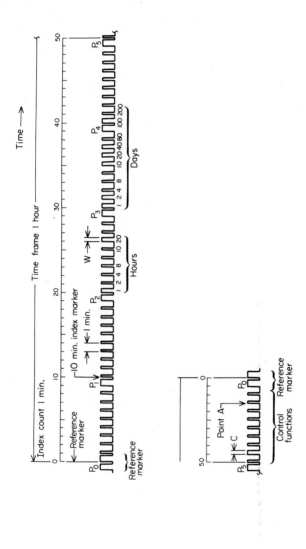

P = Position identifier, 48 sec. duration
W = Weighted code digit, 30 sec. duration
C = Control element (example), 30 sec. duration
Duration of index markers = 12 sec.

Time at point A = 21 hr, 57 min. on day 173

Fig. 4.10. IRIG standard format D 1 p.p.m.

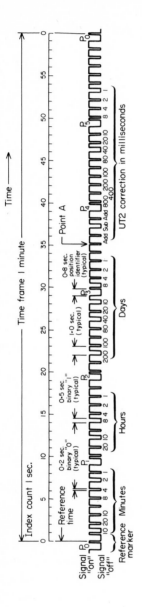

Fig. 4.11. NBS time code WWVB 1 p.p.s.

Time at point A: 258 days, 18 hours, 42 minutes, 35 seconds, subtract 41 milliseconds

Figures 4.6 to 4.11 show six* of the still most frequently used time code formats, namely NASA 36 bit, IRIG A, B, C and D and the NBS time code disseminated over the WWVB 60 kHz LF transmitter. The letters IRIG stand for Inter Range Instrumentation Group, a result of a standardization effort among US missile ranges in the early 1960s.

For transmission, the time code can be modulated on a carrier by various methods such as amplitude or frequency shift keying. One of the most frequent uses of such time codes is in data acquisition on multitrack magnetic tape recorders, whereby one track is used for the time code and the others for recording experimental data. The codes are also useful for driving many remote displays, with the advantage that no updating is required after an interruption—as with ordinary synchronous clocks or secondary clocks driven by seconds' or minutes' pulses.

With the increasing use of low-cost digital circuitry, it is possible that the use of serial time codes will become more general in the future, as most standard time dissemination services are introducing some form of time code in their signals.

REFERENCES

1. H. M. Smith. International time and frequency coordination. *Proc. IEEE*, **60** (5), 479–487 (1972).
2. J. T. Henderson. The foundation of time and frequency in various countries. *Proc. IEEE*, **60** (5), 487–493 (1972).
3. G. M. R. Winkler. Note on the use of "Epoch", "Date" and similar terms. *Proc. IEEE (Letter)*, **60** (5), 638 (1972).
4. P. Kartaschoff and J. A. Barnes. Standard time and frequency generation. *Proc. IEEE*, **60** (5), 493–501 (1972).
5. NASA Contractor Report. "Study of Methods for Synchronizing Remotely Located Clocks", prepared by Sperry Gyroscope Company. NASA CR-738, Washington, D.C., March 1967.
6. CCIR Recommendation 460 (Rev. 74), approved by the CCIR XIIIth Plenary Assembly, Geneva, 1974. Documents available from ITU Headquarters, Geneva, Switzerland.
7. J. A. Barnes. Basic concepts of precise time and frequency. *In* "Time and Frequency: Theory and Fundamentals" (B. E. Blair, Ed.), NBS Monograph 140, pp. 1–40. US Govt Printing Office, Washington, DC, May 1974.
8. J. A. Barnes. Atomic timekeeping and the statistics of precision signal generators. *Proc. IEEE*, **54** (2), 207–220 (1966).
9. D. W. Allan. Time measurement of frequency and frequency stability of precision oscillators. *In* "Proc. 6th Ann. PTTI Planning Meeting" (US Naval Research Laboratory, Washington, D.C., 3–5 December 1974). NASA Goddard Space Flight Center, Greenbelt, Md, NASA Doc. No. X-814-75-117.

* The code formats shown in Figs 4.6 to 4.11 are taken from a document prepared in 1965 for NASA by the Electronic Engineering Co. (EECO), Santa Ana, California.

10. H. Brandenberger, Groupe Etalons de Fréquence, Ebauches S.A., Neuchâtel, Switzerland. Private communication.
11. G. M. R. Winkler, R. G. Hall and D. B. Percival. The US Naval Observatory clock time reference and the performance of a sample of atomic clocks. *Metrologia*, **6**, 126–134 (1970).
12. Bureau International de l'Heure. "Rapport annuel pour 1973", pp. A7–A14. Paris, 1974.
13. B. E. Blair (Ed.). "Time and Frequency: Theory and Fundamentals", NBS Monograph 140, Chapter 9. pp. 205–231. US Govt Printing Office, Washington, DC, May 1974.
14. K. Yoshimara. "The Generation of an Accurate and Uniform Time Scale with Calibrations and Prediction", NBS Technical Note 626. US Govt Printing Office, Washington, DC., November 1972.
15. W. A. Ganter. "Modeling of Atomic Clock Performance and Detection of Abnormal Clock Behavior", NBS Technical Note 636. US Govt Printing Office, Washington, DC., March 1973.

APPENDIX 4.1

RECOMMENDATION 459

A NOTATION FOR REPORTING CLOCK READINGS AND FREQUENCY-GENERATOR VALUES

(Question 3/7)

The CCIR,   (1970)

CONSIDERING

(a)   that there exists at present considerable confusion in the practices used to express time differences between clocks;

(b)   that there is sometimes uncertainty as to the interpretation of reported times of reception relative to local clocks;

(c)   that there also exists ambiguity in the reporting of frequency differences;

(d)   that there is an urgent requirement for standardization of terminology and conventions in regard to measurements of frequency and time differences in order to avoid errors;

(e)   that the International Astronomical Union (IAU) in its fourth session of 29 August 1967 has adopted a Resolution concerning such conventions (Commission 31, Resolution No. 2) which helps satisfy requirements of clarity, preciseness, and usefulness for application in the field of radio time signals;

UNANIMOUSLY RECOMMENDS

1.   that, to avoid any confusion in the sign of a difference in indicated time

between clocks, or in frequency between frequency sources, algebraic
quantities should be given;

2.   that the following definitions and conventions may be used in con-
junction with the algebraic expressions:

2.1   the time and location of a clock reading or a frequency measurement
should always be designated;

2.2   at time T let a denote the reading of a clock A and b the reading of clock
B. The difference of the readings is $a - b$ and will be conventionally
designated

$$A - B = a - b \qquad (1)*$$

2.3   let the frequency of a frequency source C be denoted by f_C and that of a
frequency source D by f_D. Then the frequency difference is $f_C - f_D = \Delta f$
and may be conventionally designated as

$$C - D = \Delta f \qquad (2)$$

the nominal frequency of C and D should also be specified;

2.4   the fractional or relative frequency deviation of a frequency source C
from its nominal value f_{nC} is defined as

$$F_C = (f_C/f_{nC}) - 1 \qquad (3)$$

2.5   the fractional difference in frequency between two frequency sources
"H" and "K" is the difference in their fractional frequency deviations

$$S = F_H - F_K$$

and may also be designated conventionally:

$$H - K = S \qquad (4)†$$

2.6   a time comparison between a clock and a received time signal should
follow the conventions given in §§ 2.1 and 2.2 above; a frequency
comparison between an oscillator and a radio frequency emission
should follow the conventions given in §§ 2.1, 2.3, 2.4 and 2.5.

* Example: The result of a time comparison between the portable clock, P7, and the time
scale UTC of the BIH, measured at the BIH, would be reported as follows:
UTC(P7) − UTC(BIH) = −12·3 μs (7 July 1968, 14 h 35 min UTC; BIH).

    † Example: The result of a frequency comparison, related to the previous example, may be
reported conventionally, as a relative frequency difference, as follows:
P7 − BIH = +5 × 10^{-13} (7 July 1968, 14 h 35 min to 20 h 30 min UTC; BIH).

5

Frequency and Period Measurements by Means of Counters

5.1. INTRODUCTION

Counters for frequency, period and time interval measurements were introduced over twenty years ago and have since gained wide acceptance as examples of modern digital instrumentation. Most counters in use today are very flexible in their mode of operation and highly sophisticated in their internal design.

In this chapter, we shall not give a complete review of the various instruments manufactured industrially but a description of the basic principles of measurement. In most cases, the various modes of operation are combined in one instrument; some additional features are sometimes added by means of plug-in modules. The principles described in the following paragraphs are easily recognizable in the operating manuals supplied with the instruments. For many refinements and latest details in developments, the reader is to refer to the information published by manufacturers and to reviews appearing in trade journals as such details have changed and will go on changing more than most of the basic principles reviewed below.

5.2. DIRECT FREQUENCY MEASUREMENTS BY COUNTING THROUGH A TIMED GATE

This measurement method is most directly related to the concept of frequency, i.e. the number of events per unit time interval. Figure 5.1 shows

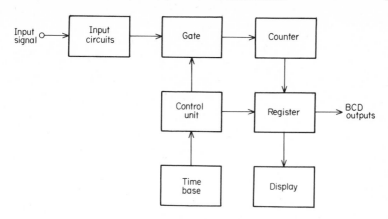

FIG. 5.1. Direct frequency measurement.

the basic block diagram of a setup including all important functions currently found in practical instruments.

In the following description of the various functional units, some details equally apply to other configurations described in later paragraphs and will therefore not be repeated therein.

5.2.1. Input circuits

The input circuits must fulfill a twofold purpose, namely adapting

(a) the amplitude; and
(b) the waveform

of the input signal to uniform values for efficient handling by the gate and the first counting unit of the counter. In order to accommodate a wide range of voltages and frequencies, the input circuits usually contain an attenuator to reduce high level signals as well as some means of overload protection, followed by a wideband d.c. coupled amplifier to handle low signals, the gain being limited only by noise considerations. A.c. coupling can sometimes be switched in to handle weak or moderate a.c. signals riding on a high d.c. level. These parts of the input circuits are usually similar to input attenuators and amplifiers found in oscilloscopes. A variable d.c. level adjustment is useful too.

The second function, namely waveform processing, may then operate at a convenient voltage level. Understanding this operation is most important for successful and accurate measurement since incorrect adjustments can lead to great errors.

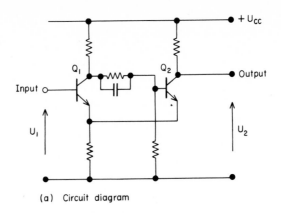

(a) Circuit diagram

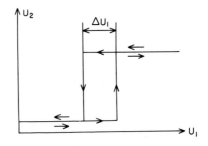

(b)   Input / output characteristics

FIG. 5.2. Principle of Schmitt-trigger circuit.

Waveform processing is usually effected by means of a "Schmitt-trigger"-circuit which is a bistable circuit shown in Fig. 5.2(a). Figure 5.2(b) shows the relationship between the input voltage U_1 and the output voltage U_2. The principle of operation is simple: if U_1 is low, Q_1 is blocked and Q_2 conducts, and therefore U_2 is also low. Increasing U_1 will bring Q_1 into conduction. Regenerative action as in a flip–flop will cause a rapid switchover so that Q_1 conducts heavily and Q_2 is blocked: U_2 jumps to a high level. The two states can be stable only if there is a hysteresis shown as ΔU_1 in Fig. 5.2 (see Ref. 43 in Chapter 3). The value of ΔU_1 is determined by the detailed circuit design. The same is true for the rise and fall times of the switching waveform which must be adapted to the counting circuits to be driven.

The counter changes state, i.e. registers one count for each positive transition (or each negative but not for both) of the U_2-pulses.

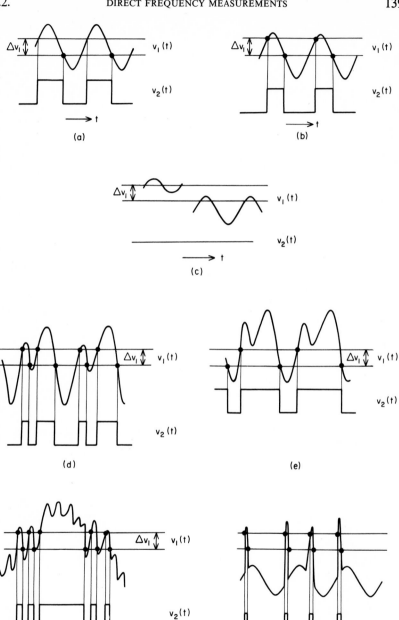

FIG. 5.3. Waveforms.

Figure 5.3 shows the action of the Schmitt-trigger circuit for various typical cases: (a) and (b) show the effect of a d.c. level shift on a sinusoidal wave. Here, only the width or duty-cycle of the pulses changes. However, if the level is outside the ΔU_1 hysteresis or "trigger window" or if the signal level is too low, switching is no longer possible as shown in (c). An example of what can happen with more complex waveforms is shown in (d) and (e). Twice as many pulses are generated in (d) than in (e), only by changing the d.c. level of the signal. If noise is superposed on the signal as in (f) and (g), the number of pulses generated no longer has any relationship with the fundamental frequency of the signal.

These few examples should suffice to illustrate that a counter is, in fact, a somewhat blind and stupid instrument which should never be used on signals having an unknown waveform, noise and unwanted components content. Therefore, the experimenter is well advised *always to use an oscilloscope alongside a counter*, at least for initial operation of any new or modified setup. In most cases it is then possible to avoid errors just by seeing the signal before attempting to measure its frequency or period. Obviously, the sensitivity and frequency range of the oscilloscope should be compatible with that of the counter. In more difficult cases, it is advantageous to have a counter in which the trigger levels and possibly the processed waveforms before the gate are available on test ports.

It will sometimes be necessary to connect appropriate low-pass or band-pass filters in front of the counter. Using shielded cables and avoiding ground-loops (mains frequency pickup) is another recommended practice.

5.2.2. Gate

The purpose of the input gate is to start and to stop the counting operation. Its function is in principle that of a conventional logic AND-gate. The design must nevertheless be arranged in such a way that only the logic transitions at the signal input port are counted and not the transitions at the on–off control port when there is no signal at the input. For very high-speed operation, special designs utilizing hybrid and thin-film circuit technology are used. The voltage levels to be handled have to be compatible with the circuit technology used in the counting unit.

5.2.3. Time-base

The time base controls all timing operations in the system. Its frequency source is usually a stable quartz crystal oscillator followed by frequency dividers to establish the required time intervals. For direct frequency measurements the most common designs provide decimal fractions and multiples of one second. The input gate is opened for a selected time interval during which the input transitions are counted. After stopping the count and if the counter started at zero, the registered count will then be a decimal fraction or multiple of the measured frequency in hertz and establishing the correct value is only a matter of placing the decimal point at the proper place.

Errors of the time base frequency have a direct influence on the displayed result. Therefore, the more expensive counters usually include a high quality oven-controlled crystal oscillator with low ageing.

In laboratories where atomic standards and a standard frequency distribution system are available, it may be advantageous to use that signal on the external time base input existing on most instruments. This often allows high precision measurements to be made with inexpensive counters.

5.2.4. Counter, register and display

The basic building block of the counter is now generally a BCD-counting unit similar to that described in Section 3.4.2 (Fig. 3.21). In modern instruments, large scale integrated circuits can be found which contain several BCD-units connected in a chain on the same chip. The maximum speed of the counter is determined by the first decimal counting unit (i.e. that next to the input gate) for which special circuits are sometimes used.

As the starting and the stopping of the counter are controlled by the input gate, the only further external control required is a reset to zero before beginning a new counting cycle.

The information registered in the stopped counter is available as logic levels on the four output terminals A, B, C, D, for each decimal digit, according to the state table (e.g. Table 3.7) of the BCD unit.

This information is transferred into a register or memory, usually consisting in another set of J–K flip–flops. The control unit shown in Fig. 5.1 must provide the logic signals making this transfer after each stop of the counter and preventing a change of state of the register during the active counting cycle. The information registered after the last counting cycle is thus retained until the end of the following cycle and available as a parallel BCD code for the display and for external use, e.g. data acquisition systems, computers, D/A converters for strip chart recording, etc.

Different types of visual displays exist. They all require appropriate decoding of the BCD output signals. The most common displays are: gas discharge tubes with either numerical or seven-segment cathodes and, more recently, light emitting diodes (LED) in seven segment or dot-matrix form. In portable instruments designed for low-power consumption, liquid-crystal displays are another possibility.

5.2.5. Control unit

The block called control unit in Fig. 5.1 represents all common function
required for correct operation of the various possible functional modes of the
counter. For direct frequency counting, it selects the desired counting time
interval and provides the required logic functions to execute the programme

(1) open gate	counting starts
(2) close gate	counting stopped
(3) transfer BCD count into register	count registered result displayed and available on BCD output lines
(4) reset counter to 0	counter ready for next cycle
(5) prepare next cycle	and waits
(6) go to 1	next measurement starts.

In very simple models and in older designs the time base contains only a
continuously running chain of decade dividers. Gate time intervals for
measurements of, for example, 1 second duration are then available only
every two seconds because of the finite time needed for the display data transfer and reset operations. The gate is open for counting only during alternate
periods of the selected time base pulse sequence, i.e. there is a dead time equal
to the measurement or sampling time. With the definitions of Chapter 2, this
means that the measurement period T is equal to 2τ.

In modern counters, the control unit also controls the time base divider
enabling a new measurement cycle immediately after the reset of the previous
count. With fast logic, the dead time can thus be reduced to the order of
microseconds whatever the value selected for the sampling time τ.

As mentioned in Chapter 2, the dead time can have some influence on the
statistics of the measurements and it is therefore necessary to know the value
of the dead time, especially if very short sampling times are used.

5.2.6. Measurement errors in direct frequency measurement

The uncertainty of direct frequency measurement is basically limited by two
factors:

(a) ± 1 count ambiguity; and
(b) time base accuracy,

according to the relation

$$v = \frac{N}{\tau + \Delta\tau} \quad (\text{Hz}) \tag{5.1}$$

where N is the displayed count, $\Delta\tau$ the timing error, and

$$\frac{\Delta v}{v} = \pm\frac{1}{v\tau} \mp \frac{N\,\Delta\tau}{v\,\tau^2}. \tag{5.2a}$$

$$y = \pm\frac{1}{v\tau} \mp \frac{N\,\Delta v_t}{v\tau\,v_t} \tag{5.2b}$$

thus

$$y = \pm\frac{1}{v\tau} \mp \frac{N}{v\tau}y_t. \tag{5.2c}$$

Δv, v, y refer to the measured frequency and Δv_t, v_t, y_t to the frequency of the time base generator.

There is thus a very simple relation between measurement uncertainty, frequency to be measured and sampling time. Figure 5.4 shows this relationship as a function of frequency and sampling time. The diagonal straight lines determine the precision limit due to the ± 1 count error. The limitation due to time base inaccuracy is indicated for two cases: $\pm 1 \times 10^{-5}$ and $\pm 1 \times 10^{-8}$. The limit due to the ± 1 count error is actually a limit of resolution. Other errors can of course be introduced by false triggering as discussed in

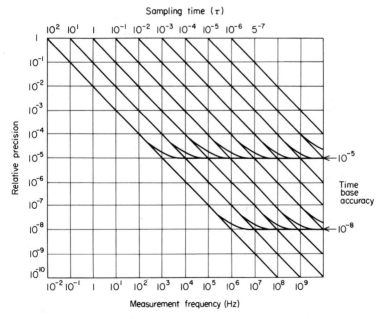

FIG. 5.4. Uncertainty of direct frequency measurement.

Section 5.2.1 (Fig. 5.3). Measurement resolution better than the limits imposed in Fig. 5.4 can be achieved by measuring the period and computing its inverse as shown in the following section.

Within the error limits mentioned above, the result of the count is the aver-times, e.g. positive zero crossings of a periodic waveform.

5.3. PERIOD MEASUREMENT

Period measurement is a special case of time interval measurement, namely the measurement of the time interval between two successive equal phase-times, e.g. positive zero crossings of a periodic waveform.

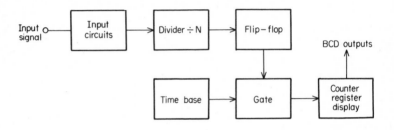

FIG. 5.5. Period measurement.

A block diagram of a period measurement setup is shown in Fig. 5.5. The input signal is first processed by input circuits as described in Section 5.2.1. The output pulses trigger a gate control flip–flop which opens the gate on the first pulse and closes it after the Nth pulse. The preset divider by N allows either single period ($N = 1$) or multiple period measurements to be made.

The timing diagrams in Fig. 5.6 illustrate the principle of measurement. When successive single period measurements are made, this arrangement implies a dead time equal to the period, i.e. only alternate periods can be measured.

In multiple period measurements, the dead-time is also equal to at least one single period, provided that a suitable control system (not shown in Fig. 5.5 for sake of clearness) restarts the cycle at the next crossing of the input trigger level.

During the opening time of the gate, the counter receives clock pulses from the time base oscillator. After closing the gate, the displayed count shows the measured time for N periods of the input signal, the units of the least significant digit being equal to the clock pulse period T_t, i.e.

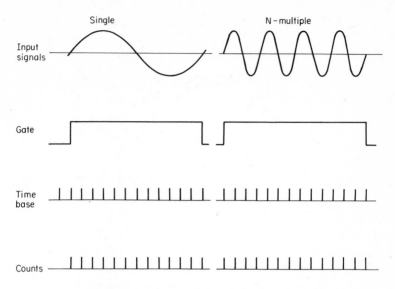

Fɪɢ. 5.6. Single and multiple period measurement timing diagrams.

$T_t = \quad 1\ \mu s$ for a $\quad 1$ MHz time base
$T_t = 100$ ns for a $\ 10$ MHz time base
$T_t = \quad 10$ ns for a 100 MHz time base, etc.

The functions of the counting units, the register and the display are the same as those described in Section 5.2. The displayed result being the duration of N periods of the input signal having the frequency v_1 and period T_1, it is necessary to do some computations in order to obtain the value of v_1, *averaged over the sampling time* $\tau = NT_1$.
The displayed result is

$$T_d = \tau(1 + y_t) = NT_1(1 + y_t) \tag{5.3}$$

where

$$y_t = \frac{v_t - v_{t_0}}{v_{t_0}}$$

is the normalized offset of the time base frequency from its nominal value. We have thus

$$v_1 = \frac{1}{T_1} = \frac{N(1 + y_t)}{T_d}. \tag{5.4}$$

This includes the measurement error due to the time base inaccuracy. The

F

error due to the resolution limited by the ± 1 count ambiguity shows up in the displayed result as

$$\Delta T_d = \pm T_t$$

T_t being the period of the time base clock pulse frequency, higher order errors due to $y_t \neq 0$ being negligible here.

Including this in (5.4), we then have

$$v_1 + \Delta v_1 = \frac{N(1 + y_t)}{T_d + \Delta T_d}$$

which, neglecting higher order terms reduces to the following expression for the error Δv_1

$$\Delta v_1 = \frac{N}{T_d} y_t - \frac{N T_t}{T_d^2}. \qquad (5.5)$$

the first term being due to the time base inaccuracy and the second to the ± 1 count ambiguity. There is however another source of error occurring typically in period and time interval measurements but not in direct frequency measurements, namely trigger errors due to noise, spurious signals and d.c. level variations as shown in Fig. 5.3.

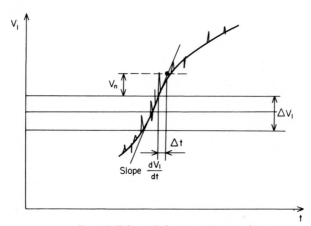

FIG. 5.7. Trigger timing error due to noise.

The timing error due to a spurious peak noise voltage $\pm U_n$ is directly related to the slope of the input waveform $u_1(t)$ at the switching level of the trigger window, as illustrated in Fig. 5.7:

$$\Delta t = \pm \frac{U_n}{du_1/dt}. \qquad (5.6)$$

Combining the influence of all three errors on the measurement of v_1, we obtain:

$$\Delta v_1 = \frac{N}{T_d} y_t - \frac{N}{T_d^2}(T_t + \Delta t_1 + \Delta t_2) \tag{5.7}$$

Again, higher order terms have been neglected since all error terms must be small in order to allow meaningful measurements to be made.

Δt_1 and Δt_2 represent the trigger errors occurring at the beginning and at the end of measurement interval. The normalized error $y_1 = \Delta v_1/v_1$ is estimated in the worst case, assuming that all errors add by

$$y_1 = y_t + \frac{T_t + \Delta t_1 + \Delta t_2}{T_d} \tag{5.8}$$

This shows that for short period measurements and high time-base frequency, i.e. T_d and T_t small, the trigger errors become the dominant factor affecting the accuracy of measurement. The advantage of period measurement becomes obvious especially at low input frequencies where the precision of direct frequency measurement is unduly limited by the ± 1 count ambiguity as shown in Fig. 5.4. Furthermore, the minimum time required to make a period measurement with a given resolution is much shorter than that required for a direct frequency measurement. Some recent instruments can display frequency measured by means of a period measurement, whereby the necessary computation is effected by a built-in microprocessor.

5.4. FREQUENCY RATIO MEASUREMENTS

If we replace the time base in Fig. 5.5 by another input unit as shown in Fig. 5.8, it is possible to measure the ratio of the two frequencies of the signals applied at the inputs A and B. According to the timing diagram of Fig. 5.9,

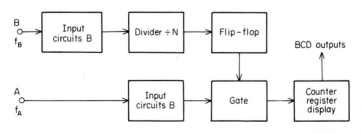

FIG. 5.8. Frequency ratio measurement.

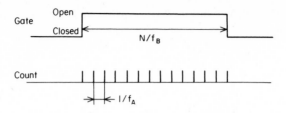

FIG. 5.9. Frequency ratio measurement.

the gate is open during the time interval N/f_B. The period of the signal accumulated in the counter is $1/f_A$, thus the displayed count N_d is equal to

$$N_d = N \frac{f_A}{f_B}. \tag{5.9}$$

The resolution of this measurement depends on the ± 1 count ambiguity of f_A, and it is therefore recommended to apply the higher of the two frequencies to input A. The influence of trigger error must be determined for each particular case according to the considerations given in the preceding Section 5.3.

5.5. TIME INTERVAL MEASUREMENTS

Time interval measurements by means of counters require two independent input channels A and B as shown in Fig. 5.10. Channel A is the "start" and B

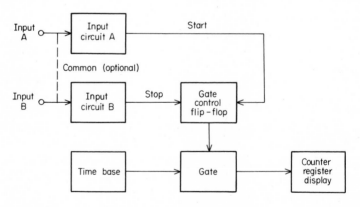

FIG. 5.10. Time interval measurement.

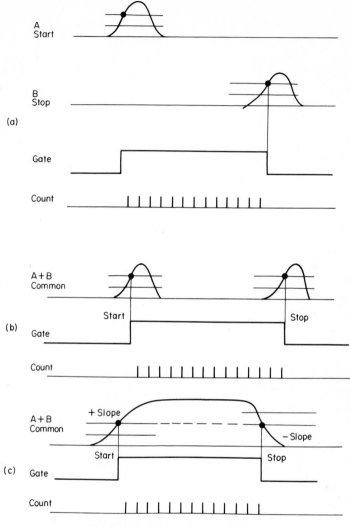

FIG. 5.11. Time interval measurements.

the "stop" channel. These channels are fed into the set and reset inputs of the gate control flip–flop so that the system operates as follows:

(1) If the gate is closed and the counter reset to zero, the next pulse in channel A opens the gate by setting the gate control flip–flop. The counter starts accumulating clock pulses from the time base. Further pulses in the A-channel have no effect.

(2) A pulse in the B-channel resets the gate control flip–flop and closes the gate. The accumulated count is stored and displayed as usual. Further pulses in the B-channel have no effect. At the end of the storage and display interval, the counter is reset to zero. Control logic not shown in Fig. 5.10 must inhibit channel A until the completion of the reset to zero operation preventing the start of a new measurement before completion of the previous one.

Channels A and B can either be operated independently as shown in Fig. 5.10 by the solid lines or in common as shown by the dotted line. Further illustrations of possible modes of operation are given by the timing diagrams in Fig. 5.11. In (a), we have a measurement of the time interval between two pulses arriving on the separate channels A and B, whereby the trigger points are set at selected levels on the positive slope of the pulses. In (b), we measure the time interval between two successive pulses from a common source. The trigger points are again set on the positive slope. For this mode to operate correctly, the gate control flip-flop must operate in a toggling mode: since it receives logic signals on both inputs simultaneously, it must change state each time it receives the logic level transitions from A and B.

In (c), we measure the width of a pulse. Here, channel A is set to trigger on the positive slope and B on the negative slope, the trigger windows being shifted so that the triggering occurs at the same level at both sides of the pulse. In some instruments, this trigger window shift is done automatically.

These are only three very elementary examples of time interval measurement and it is easy to elaborate much further on many special and more sophisticated cases. What appears again clearly from Fig. 5.11, is the great importance of *correct trigger level adjustment*. We do not measure just time intervals between "pulses" but time intervals between instants where the input waveforms cross some predetermined voltage levels.

The considerations on measurement errors discussed in Section 5.3 apply also to time interval measurement, mainly those on triggering error due to noise (Eq. (5.6)).

With modern counters using time base clock frequencies up to 500 MHz a resolution of ± 2 ns in single time interval measurements can be obtained. Systematic errors due to different delays in the channels A and B must therefore be taken into account and corrected. One nanosecond being equivalent to approximately 20 cm of coaxial cable, the experimental setups involving fast time interval must be laid out with care.

In measurements involving more complex waveforms and pulse sequences, it is not always easy to see which time interval is actually measured by the counter. In this case, the use of oscilloscopes is again strongly recommended. Using the gate pulse to intensify the oscilloscope trace during the counting interval is a useful technique for the identification of the measured interval.

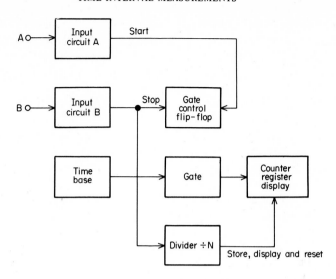

FIG. 5.12. Time interval averaging.

Triggered pulse generators with adjustable delay are used in many applications to inhibit the gate during part of the measurement cycle, so waiting for the desired pulse and preventing unwanted pulses from changing the state of the gate control flip–flop.

On repetitive waveforms, the resolution of time interval measurement can be improved by means of *time interval averaging*.

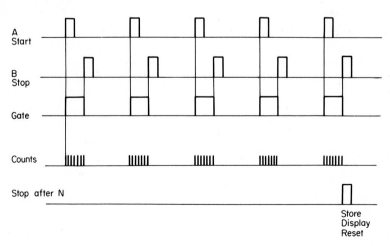

FIG. 5.13. Time interval averaging.

The principle of this mode is illustrated by the block diagram of Fig. 5.12 and the timing diagram of Fig. 5.13. The stop pulses from channel B do not actuate the storage-display-reset operation in the counter directly but through a preset $\div N$ divider. The results of N successive counts are thus accumulated in the counter and only the sum is stored and displayed after N measurements. If N is a power of ten, say 10, 100, 1000, etc., the average is easily displayed by proper placing of the decimal point.

In order to obtain a statistical average of the individual measurement, the repetition period of the measurements must not be synchronous with the time base. If the statistics of the measured individual intervals can be described by a white noise process, the resolution due to time interval averaging improves proportionally to $N^{-\frac{1}{2}}$. Thus, with a 500 MHz counter and averaging of $N = 10000$ intervals, a resolution of 20 ps appears to be feasible.

5.6. INCREASING THE RESOLUTION OF FREQUENCY MEASUREMENTS

As shown in the preceding sections, the resolution of frequency and period measurements is limited by the ± 1 count ambiguity, especially if measurements are to be made within limited measurement time intervals. The classical method to increase the resolution of frequency measurement is to measure a beat frequency produced by mixing a harmonic of the unknown signal with a harmonic of a standard frequency generator.

The principle of this technique, also called the heterodyne method, is shown in Fig. 5.14. The frequency v_1 of the unknown is multiplied by a factor n_1 and the standard frequency v_0 by a factor n_0. The factors n_1 and n_0 are chosen so that the beat frequency, i.e., the difference $|n_1 v_1 - n_0 v_0|$ is low compared to the harmonic frequencies $n_1 v_1$ and $n_0 v_0$ applied to the mixer inputs. This condition restricts somewhat the possible ranges of frequencies

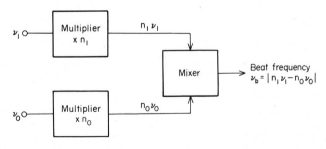

FIG. 5.14. Heterodyne method.

v_1 and multiplication factors n_1 and n_0 but in many cases a common multiple can be found. If this should not be the case, some form of synthesizer can be used (see Section 3.5).

As the mixer does not reproduce the sign of the difference $n_1 v_1 - n_0 v_0$, it is important to make sure that it is known beforehand and remains the same throughout the experiments. No possible variation of v_1 should drive v_b through "zero beat". Violation of this rule leads to gross errors, especially if the signal to be measured is frequency-modulated.

The desired increase in resolution by using the heterodyne method is obtained by means of period measurement on the beat frequency v_b. By measuring N periods, the displayed result is

$$T_d = \frac{N}{v_b}.$$ (5.10)

The time base period error is $\pm \Delta T_d$ and the ± 1 count ambiguity leads to a resolution of

$$\Delta v_b = \mp \frac{N}{T_d^2} \Delta T_d$$ (5.11)

on the beat frequency v_b. Expressed as normalized resolution on the measured frequency v_1, we have:

$$\frac{\Delta v_1}{v_1} = y_1 = \mp \frac{N}{n_1 v_1 T_d^2} \Delta T_d.$$ (5.12)

The effect of increased resolution due to heterodyning is best illustrated by a numerical example.

Let the frequency to be measured be:

$$v_1 = 9750 \text{ kHz}.$$

We have a counter allowing direct frequency and multiple period measurement with N variable between 1 and 10^6 periods in unit steps. We wish to take measurements, each lasting for one second. The time base frequency is 10 MHz and assumed to be accurate. The *direct frequency* measurement with 1 second averaging time has a resolution of ± 1 cycle, i.e. a normalized resolution of

$$y_1 = \frac{\pm 1}{9 \cdot 75 \times 10^6} = \pm 1 \cdot 026 \times 10^{-7}.$$

Direct heterodyning without frequency multiplication, against the 10 MHz reference frequency v_0 yields a beat frequency

$$v_b = 10000 - 9750 = 250 \text{ kHz}.$$

For a period measurement averaged over 1 second, we adjust $N = 2 \cdot 5 \times 10^5$. The time base period is $\Delta T_d = 10^{-7}$ s, thus

$$y_1 = \mp \frac{2 \cdot 5 \times 10^5}{9 \cdot 75 \times 10^6} \times 10^{-7} = \mp 2 \cdot 56 \times 10^{-9}.$$

The resolution is improved by a factor of about 40. If we multiply v_1 by 39,

$$\text{i.e.} \, n_1 v_1 = 380 \cdot 25 \, \text{MHz}$$

and v_0 by 38,

$$n_0 v_0 = 380 \, \text{MHz}$$

the beat frequency is again

$$v_b = n_1 v_1 - n_0 v_0 = 250 \, \text{kHz}$$

but the resolution is

$$y_1 = \frac{2 \cdot 5 \times 10^5}{39 \times 9 \cdot 75 \times 10^6} \times 10^{-7} = 6 \cdot 57 \times 10^{-11}$$

for the same averaging time of 1 second, i.e. again an improvement of 39 (which is the multiplication factor n_1) with respect to direct heterodyning and of 1560 with respect to direct frequency measurement.

It is also worth noting that in the above given example, an error in the time base frequency of 1×10^{-7} changes the result only by one times y_1 whereas any error of the reference frequency $v_0 = 10$ MHz shows up directly in the result.

From Eq.(5.12) it can be seen that a high resolution is obtained if the beat frequency is low and the factor n_1 large. The limits of this method are given only by the noise enhancement due to the frequency multiplication.

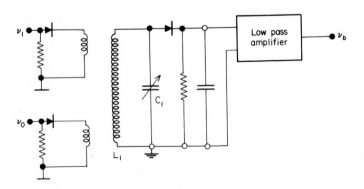

FIG. 5.15. Simple harmonic mixer circuit.

A very simple circuit which is useful for laboratory experiments is shown in Fig. 5.15. It can be used throughout the HF and VHF range and is composed of two diode harmonic generators coupled to a resonant LC-circuit tuned to the desired common multiple frequency. The required input power level is approximately 1 V RMS on 50 Ohms.

Gold doped germanium diodes were used in the original version in 1959, but any fast diode capable of withstanding the applied r.f.-power will probably work.

5.7. MICROWAVE FREQUENCY MEASUREMENTS

The maximum counting speed of a counter is limited by the available technology of active switching devices. The current state of the art allows a maximum frequency of more than 500 MHz in the standard range of available commercial instruments. Special devices working above 1 GHz have been developed more recently and it is certain that the trend to faster logic that will handle still higher frequencies will continue. There are also methods for measuring frequencies in the microwave region by means of counters limited to 500 MHz or less.

With the exception of fast pre-scalers, i.e. frequency dividers placed in front of the counter, there are two basic methods in current use, namely

(a) heterodyne converters; and
(b) transfer oscillators.

The principle of *heterodyne converters* is similar to that discussed in the previous section, except that only the reference frequency is to be multiplied into the range of measurement. Figure 5.16 illustrates this principle. The time base generates a reference frequency which is multiplied in a first stage to a value v_0 which is somewhat lower than the maximum counter frequency, e.g. $v_0 = 200$ MHz. This signal drives a step recovery diode "comb generator" (see Section 3.4.1) which, by virtue of its sharp narrow pulses creates a spectrum with regularly spaced signals at $n \times (v_0 = 200$ MHz$)$.

A tunable filter is then used to select a desired harmonic which is mixed with the unknown signal of frequency v_1. The beat frequency

$$v_b = v_1 \pm nv_0 > 0 \qquad (5.13)$$

is amplified by the low-pass video amplifier and fed into the counter where its frequency is measured, stored and displayed in the usual way. The measured frequency is therefore equal to

$$v_1 = nv_0 \pm v_b. \qquad (5.14)$$

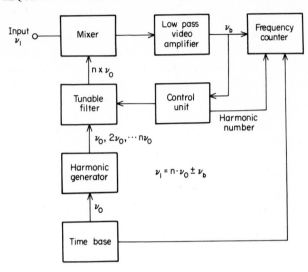

FIG. 5.16. Heterodyne converter.

The problem is the ambiguity inherent in the heterodyning method: depending on the selected harmonic number, the result of the count must be either added to or subtracted from the selected reference harmonic $n\nu_0$. The ambiguity can be resolved by systematically approaching the unknown signal from lower frequencies.

In modern instruments, the harmonic selection filter is based on the ferromagnetic resonance property of yttrium–iron–garnet (YIG-filter) which allows the design of electronically tunable narrow band filters operating over a wide range of frequencies from VHF up to 20 GHz. Automatic operation is thus possible by means of a control unit which sweeps the filter upwards through the operating range until a beat is detected. The sweep is then stopped, the counter started and the selected harmonic number automatically fed into the display.

The advantage of the heterodyne method, especially if automatic, is that the resolution is the same as if the high frequency ν_1 were measured directly. Furthermore, because of the wide-band characteristics of the video-amplifier, the counter is relatively tolerant of frequency-modulated input signals. The major drawback, at least in early models, was that the sensitivity, i.e. the minimum signal level required for correct operation was not as good as with the transfer-oscillator method.

The *transfer oscillator* method is illustrated by Fig. 5.17. Here, the unknown input signal of frequency ν_1 is mixed with the nth harmonic of a variable

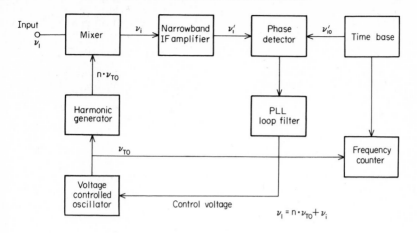

FIG. 5.17. Phase-locked transfer oscillator.

frequency voltage controlled oscillator (VCO) to produce an intermediate frequency v_i of eg. 1 MHz. v_i is compared to its nominal value v_{i0} derived from the time base in a phase detector. The resulting output voltage is filtered and fed into the control voltage input of the VCO, thus achieving a phase-locked loop (PLL). The basic relation of the frequencies is:

$$v_1 = nv_{TO} + v_i.$$

v_{TO} is measured by the counter, v_i is known ($v_i = v_{i0}$ if PLL) and therefore v_1 can be computed if the harmonic number n is known. n can be determined by locking the VCO on two successive harmonics:

$$v_1 = nv_{TO1} + v_i \qquad \text{first lock}$$
$$v_1 = (n - 1)v_{TO2} + v_i \qquad \text{second lock}$$

then

$$n = \frac{v_{TO2}}{v_{TO2} - v_{TO1}}.$$

The operation of a transfer oscillator system can also be made automatic but the detailed design is more complex than with the heterodyne converter. The main advantage of the transfer oscillator method is its potential higher sensitivity due to the good signal-to-noise ratio obtained through the narrowband i.f. amplifier and PLL circuit. However, there are some disadvantages in comparison with the heterodyne converter, e.g.

the *resolution* is reduced by the fact of n when v_{TO} is measured directly by the counter;

the *tolerance for FM* is greatly reduced since the PLL might lose the lock on the carrier of the input signal and lock on a sideband instead, producing erroneous results.

For the above given reasons, for most applications, heterodyne converters with YIG filters are presently preferred to the older transfer oscillator technique.

REFERENCES

This chapter has been compiled from various operating manuals, application notes and data sheets from the following instrument manufacturing companies:

EIP Inc., Santa Clara, California
Hewlett-Packard Co., Palo Alto, California
Philips, Eindhoven, Netherlands

6

Phase-time Measurements

Phase-time measurement techniques have their most interesting and useful application in frequency and time comparisons between very stable oscillators and clocks over medium and long time intervals, where the frequencies of the oscillators to be compared are almost equal. These techniques are complementary to those using counters as discussed in the preceding chapter. They are less universal but have the advantage of being reliable and inexpensive.

The oldest phase comparison method for measuring the duration of a beat period between oscillators of almost equal frequency is the Lissajous-pattern observed on an oscilloscope where the two signals to be compared are connected to the y and x axis inputs respectively. If the two signals are equal amplitude sinusoids, the pattern is an ellipse degenerating into a straight line when the signals are exactly in phase. Repetition of this condition can easily be measured by means of a stop-watch. Two 1-MHz signals having a relative offset of 10^{-8} reproduce this condition every 100 seconds. With the stop-watch, a skilled operator achieves a precision of about ± 0.1 s, i.e. he can measure the frequency offset with a precision of $\pm 1 \times 10^{-11}$ on the averaging time of 100 s. This is actually equivalent to a period measurement of a slow beat frequency.

It is also possible to make phase-time difference measurements using a counter in the time-interval mode, as shown in Section 5.5. For many applications, phase comparators generating a d.c. voltage proportional to the phase difference are entirely satisfactory and much less expensive. The basic ideas leading to the design of linear phase comparators are illustrated in Fig. 6.1. First, the two sinusoidal input signals are transformed into square waves (Fig. 6.1a–d). The most rudimentary form of a linear phase comparator is obtained by combining the two square waves in a logic AND-gate. The output

159

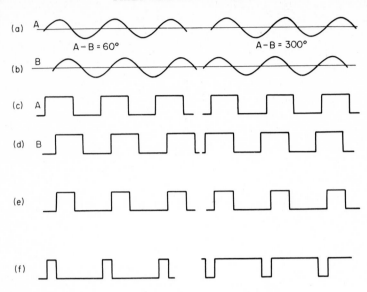

FIG. 6.1. Phase comparator waveforms.

(a), (b)   Sinusoidal inputs.
(c), (d)   Square waves, transformed inputs.
(e)   Output of AND-gate 180° linear phase comparator before low-pass filter.
(f)   Output of linear unambiguous 360° phase comparator before low-pass filter.

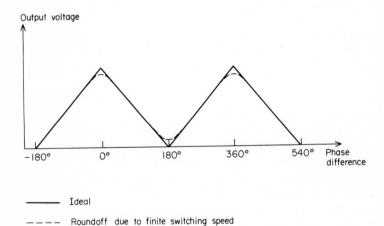

FIG. 6.2. Output voltage of 180° linear phase comparator.

of the gate is high only if both inputs are high (Fig. 6.1e). The result is a sequence of pulses having a duration proportional to the phase difference. The d.c. component of the pulse sequence, obtained by feeding it through a low-pass filter, is proportional to the time average of the pulses, thus also proportional to the phase difference. This voltage is maximum when the two input waves are in phase and zero when they are out of phase.

Figure 6.2 shows the response as a function of phase difference. The linearity is as good as the approximation of ideal square waveforms. At very high frequencies, the limited rise-time causes a roundoff of the response at the limits of the output voltage range as shown by dotted lines.

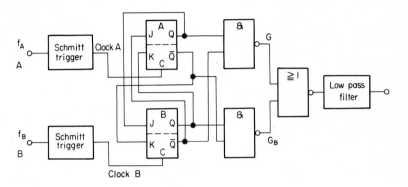

Fig. 6.3. Linear 360° phase comparator.

This simple form of phase comparator is often used in phase-locked loop circuits. For measurement purposes, its major disadvantage is the impossibility of telling whether the phase of A leads that of B or vice versa. If the frequencies of A and B are slightly different, it is likewise not possible to tell which one is the greater. Furthermore, the range of linear phase to voltage conversion is only 180° although it could be changed to 360° and almost ideal squarewaves generated by using divide-by-two flip–flops at each input; the sign-ambiguity of the phase difference nevertheless remains.

It is possible to build linear 360° phase comparators by means of suitably programmed J–K flip–flops. The desired logic output waveform is that of Fig. 6.1f, where the leading edge of A switches the output from 0 to 1 and the leading edge of B switches the output back to zero. Examples are given in Refs. 1 and 2. Another possibility depending less on special types of integrated circuits is shown in Fig. 6.3. This phase comparator uses two J–K flip–flops, two NAND-gates and one NOR-gate. The timing diagram of the comparator

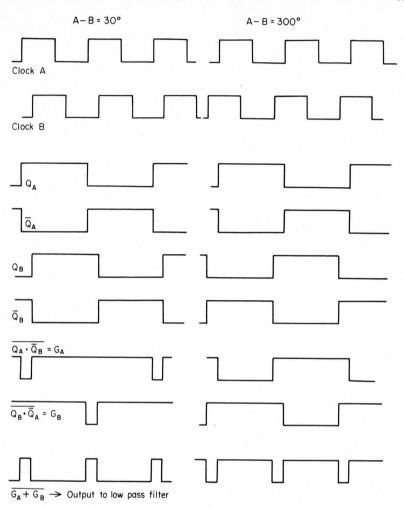

FIG. 6.4. Linear 360° phase comparator timing diagram.

(Fig. 6.4) can easily be derived using the J–K flip–flop excitation table (Chapter 3, Table 3.6). The response of the comparator at the output of the low-pass filter is shown in Fig. 6.5. The flip–flops operate as dividers by two, thus enabling the circuit to operate over a 360° range. The connections of the J and K inputs determine the priority of input A to switch the output always from 0 to 1, whereas the input B can only switch the output from 1 to 0. Ideally, the transition from 360° to 0° should take place instantly without the slight delay shown in Fig. 6.5. In practice, especially for input frequencies approaching the speed limit of the integrated circuits used, that is not possible because of the finite time delays required between the setting of the J and K inputs and the arrival of the clock pulse. With TTL-logic, the uncertainty in the transition from 360° to 0° reduces the total range of measurement only by a few percent, which can be tolerated for most applications. The advantage of

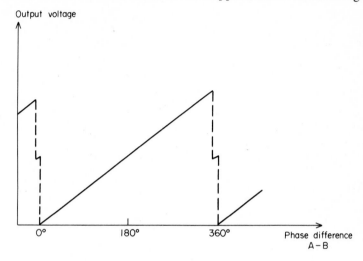

FIG. 6.5. Linear 360° phase comparator output voltage.

these 360° phase comparators compared to the 180° devices mentioned before is the definite indication of the sign of a frequency difference. In the configuration of Fig. 6.3, the output voltage ramp has a positive slope if $f_A > f_B$ and a negative slope if $f_A < f_B$.

Input frequencies of 100 kH and 1 MHz are particularly useful for phase-time comparisons since the full-scale values of 10 and 1 microsecond are most practical for further computations. Accumulated phase-time differences can easily be determined by counting the sawtooth cycles on a strip-chart record of the output voltage.

The phase-time method of comparison is especially useful for long-term comparisons of very stable frequencies, e.g. from atomic or high quality quartz crystal frequency sources. Comparisons of this kind are therefore also used in LF and VLF receivers (see Chapter 7). Examples of computations on phase-time records are given in Chapter 4.

An alternative method for measuring phase-time differences has recently been published by the NBS.[3] It uses two mixers and a transfer oscillator to

produce two beat signals from two oscillators having nominally equal frequencies. A time interval counter is then used to measure the time difference
between positive zero crossings of the two beat signals. In the first order the
phase fluctuations of the transfer oscillator cancel out. A block diagram of
this setup is shown in Fig. 6.6. Its main advantage is the very high time resolution which can be obtained (below 1 picosecond).

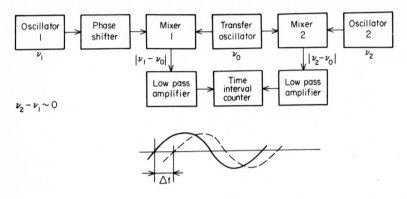

Fig. 6.6. Dual mixer phase-time difference measurement.

REFERENCES

1. J. C. Hager Jr. Edge triggered flip–flops make 360° phase meter. *Electronics*, **48** (17), 100–101 (1975).
2. C. A. Herbst. Detector measures phase over full 360° range. *Electronics*, **44** (15), 109 (1971).
3. D. W. Allan. The measurement of frequency and frequency stability of precision oscillators. *In* "Proc. 6th Ann. PTTI Planning Meeting" (US Naval Research Laboratory, Washington, DC, 3–5 December 1974). 109–141. NASA Goddard Space Flight Center, Greenbelt, Md, NASA Doc. No. X-814-75-117.

7

Frequency Domain
Measurement Techniques

7.1. INTRODUCTION

In this chapter techniques are reviewed for the measurement of spectral densities as defined and discussed in Chapter 2. Special emphasis is given to the methods and techniques required for the investigation of very stable and clean signals, where higher resolution and sensitivity than those achievable with well-known and established methods are required. The limits of direct spectrum analysis discussed in the following Section 7.2. can be overcome by using phase-locked loop techniques originally developed at the US National Bureau of Standards. These techniques are reviewed in Section 7.3. Digital instruments like samplers, digitizers and FFT processors are not discussed herein since this class of intruments is actually much more expensive than analogue instruments and therefore better suited to the analysis of very complex signals and noise phenomena in fields outside the scope of this book. Some basic principles and problems related to discrete frequency-domain analysis have been discussed in Section 2.4.

7.2. DIRECT SPECTRUM ANALYSIS

The basic method of analogue spectrum analysis is to send the signal through a band-pass filter having a centre frequency adjustable over the range in which the analysis is to be made. The bandwidth and shape of the response of the filter determines the resolution by which the various components of the signal spectrum can be determined.

In practice, the analysing filter operates at a fixed frequency as tunable filters are difficult to build. Spectrum analysers are built according to similar principles as superheterodyne radio receivers and incorporate the careful design features required for a precision instrument. Currently, spectrum analysers available from instrument manufacturers can be divided into two classes. Instruments having a wide range of input frequencies, moderate analysing bandwidth and built-in cathode ray tube display are known under the generic name of spectrum analyser and very widely used in telecommunications, radar and similar applications. Low frequency instruments allowing very high resolution analysis have been in use for about 40 years and are traditionally known as "wave analysers". Their field of application covers acoustics, mechanical vibration analysis, etc. The application of such a low frequency wave analyser will be shown in the following Section 7.3.

The basic building blocks of a spectrum analyser are shown in Fig. 7.1. The input signal is first adjusted to a suitable level by a calibrated input attenuator. In very wide range instruments operating in the microwave region, a preselector in the form of an electronically tuned YIG-filter is connected in front of the mixer. The purpose of the preselector is to suppress the image frequency response which is generated by the mixer: if the constant intermediate f_i is lower than the signal and local oscillator frequencies f_s and f_L the mixer produces f_i either by

$$f_i = f_{s1} - f_L, \qquad f_{s1} > f_L$$

or

$$f_i = f_L - f_{s2}, \qquad f_{s2} < f_L$$

i.e.

$$f_{s1,2} = f_L \pm f_i \qquad f_i \ll f_L, f_s.$$

In lower frequency spectrum analysers this problem can be avoided by means of up-conversion, i.e. having

$$f_L > 2f_{s(max)}$$
$$f_i = f_s + f_L.$$

In this case, the image frequency falls outside the input frequency range and it is sufficient to protect the input from unwanted signals by means of a low-pass filter.

The input signal frequency range is covered by sweeping the local oscillator frequency which is voltage controlled by the sweep generator. The latter also drives the x-axis input of the CRT display. If a preselector filter is used, its centre frequency is controlled by the same sweep generator, and precise tracking of the local oscillator and preselector frequencies is a prerequisite.

The use of YIG resonators having a linear frequency vs. current control characteristic as frequency determining elements in the preselector filter and in the local oscillator allows good tracking between these two tuned elements to be obtained in microwave spectrum analysers.

The analyser resolution bandwidth is determined by a fixed frequency filter in the intermediate frequency (i.f.) channel.

In some instruments another down-converter is used, allowing operation on a lower i.f. for which narrowband filters for high resolution analysis are easier to design.

The resolution and the sensitivity of a spectrum analyser are its abilities to separate closely spaced discrete spectral components and to extract weak discrete signals from the noise background. Both improve with decreasing analysis bandwidth.

However, the narrower the bandwidth of the analysing filter, the longer is its transient response and this puts a limit to the maximum allowable sweep or scanning speed. An approximate rule for this limit, consistent with the uncertainty relation of Eq. (2.39) can be stated as follows:

The time to scan a discrete signal through the bandwidth of the filter must be larger than the inverse value of this bandwidth. Thus, for a bandwidth of, for example, 1 kHz, the maximum sweep speed should be less than 1 kHz per millisecond or 1 MHz per second, etc.

In some instruments, the sweep speed and range switches are automatically interlocked with the resolution bandwidth selector to prevent incorrect settings. A violation of this rule would result in a distorted display, showing erroneous amplitudes of the spectral components.

There is thus a tradeoff to be made between desired resolution, total scanning frequency range and the time required to analyse a given spectral frequency range.

If the signals to be analysed are not only of a complex structure but also have components which vary with time in a more or less random manner, reliable information concerning the spectrum can only be obtained if at all by averaging a number of sample spectra.

In some recent instruments this is possible by means of a built-in digitizer system feeding a memory. Developments combining analogue spectrum analysers with computer systems have been proliferating for some time and a complete review of this rapidly developing field is beyond the scope of this text.

A most important characteristic of spectrum analysers, beside those already mentioned, is the *dynamic range* of the instrument.

The dynamic range, usually given in decibels, defines the allowed input signal level limits, which result in a display free of spurious responses. That there must be a limit, especially on high level input signals, can easily be seen from the block diagram of Fig. 7.1 by recalling that the mixer is a nonlinear element which can generate higher order intermodulation products (see also Section 3.5 and the references cited therein).

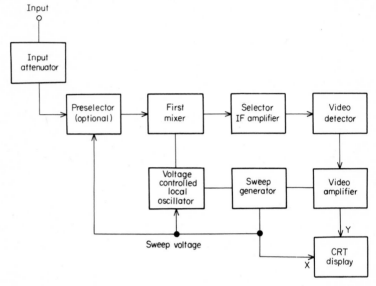

Fig. 7.1. Spectrum analyser block diagram.

Therefore, conditions can exist under which the instrument displays spectral components which are not present in the original input signal but generated within the instrument. The most critical element is usually the first mixer. In current technology using, for example, Schottky-diode double balanced mixers, the input signal power should not exceed about 1 mW. In good instruments, spurious responses can be held at about −80 dB below that level, so that the dynamic range is approximately 80 dB. This is a very large range indeed and to use it fully, modern spectrum analysers have a logarithmic display calibrated in decibels.

If during practical work an observed signal is suspected of being spurious, one simple means of checking its nature is to vary the input attenuator setting. If the observed component does not vary in amplitude according to the attenuator calibration, it is suspect. The best way to avoid spurious responses is to attenuate the input signal as much as possible so long as the desired signals can be extracted from the instrument's own noise level.

For the investigation of residual noise in very stable and spectrally pure signals, the direct spectral analysis method is not applicable because of the limited dynamic range of the usual instrument not designed for this purpose. Special methods developed during the past few years enable especially low phase noise levels to be measured. These methods using the phase-locked loop principle are discussed in the following section.

7.3. MEASUREMENT OF LOW LEVEL PHASE NOISE BY MEANS OF A PHASE-LOCKED LOOP

The phase-locked loop method for the measurement of low level phase noise described in this section has been developed in recent years at the US National Bureau of Standards,[1] for sources with the same nominal frequency. Fig. 7.2 shows the configuration of the measurement system. The output signals of

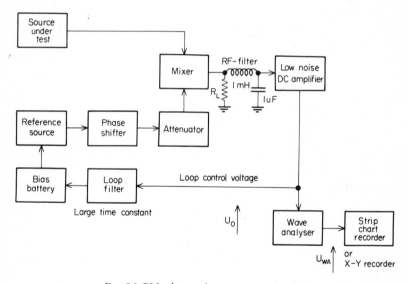

FIG. 7.2. PLL phase-noise measurement system.

the source being tested and that of the reference source are connected to the inputs of a Schottky-diode double balanced mixer. The amplitude of one of the two sources is attenuated (in Fig. 7.2 it is the reference signal) in order to ascertain that the mixer operates in the mode shown in Fig. 3.25(b)). The function of the phase shifter is described later. The two levels are adjusted so that the signal-to-noise ratio of the mixer output is optimum, i.e. the larger of the two signals should be close to the rated maximum power of the mixer. If there is no closed phase-locked loop and if the two frequencies are slightly different, the output signal must be a sinusoidal beat frequency voltage. This open loop offset condition is used for the calibration of the system as will be shown later.

The mixer output voltage is first filtered to remove the r.f.-components and then amplified by means of a low-noise d.c. amplifier. The output voltage of the amplifier is fed back to the voltage control input of the reference source through a loop filter and a bias source (adjustable battery). The loop filter is a single pole RC low-pass filter having the time constant $T_1 = RC$. The bias source is required only if the reference oscillator has no built-in bias source for its tuning varactor which must operate in the most linear part of its operating range. The output phase shifter on the reference oscillator is needed in order to bring the reference and test object signals into quadrature (i.e. 90° phase difference) as closely as possible. This makes the system insensitive to amplitude noise and the mixer output voltage a linear function of the phase difference fluctuations between the two input signals to the mixer. In exact quadrature, the mixer output voltage is zero (d.c. value) except for the noise which is to be analysed.

Let $\Delta\phi_{12}$ be the small random phase difference between the two signals. The loop keeps the long term ($\tau > 10$ s) average of $\Delta\phi_{12}$ equal to zero.

The output voltage U_0 of the amplifier is

$$U_0 = K_d \Delta\phi_{12} \quad \text{(volts)} \tag{7.1}$$

where K_d is the transfer gain of the mixer and amplifier in volts/radian.

According to the phase-lock loop equations (see Section 3.5), the power spectral density S_{U_0} of the amplifier output voltage is given by the relation:

$$S_{U_0}(f) = S_{\phi_{12}}(f)|H_l(2\pi i f)|^2 K_d^2 + S_0(f) \tag{7.2}$$

where

$$|H_l(2\pi i f)|^2 = \frac{(2\pi f)^4 T_1^2 + (2\pi f)^2}{(K_0 - (2\pi f)^2 T_1)^2 + (2\pi f)^2} \tag{7.3}$$

is the loop transfer function and $K_0 = K_2 K_d$ the d.c. loop gain of the phase-locked loop as defined in Section 3.5.

$S_{\phi_{12}}$ is the spectral density of the phase difference fluctuations $\Delta\phi_{12}$ and thus includes the phase noises of both oscillators. $S_0(f)$ is the residual noise voltage spectral density due to mixer and d.c. amplifier noise.

We now can distinguish three cases among which two are interesting:

Let S_{ϕ_1} be the phase noise spectral density of the source under test and S_{ϕ_2} that of the reference source.

If $S_{\phi_2} \gg S_{\phi_1}$, then $S_{\phi_{12}} \approx S_{\phi_2}$, i.e. we measure the phase noise of the reference and not that of the source under test. This result is not interesting, we need a better reference.

If $S_{\phi_2} \approx S_{\phi_1}$, then $S_{\phi_1} \approx \frac{1}{2} S_{\phi_{12}}$.
If $S_{\phi_2} \ll S_{\phi_1}$, then $S_{\phi_1} = S_{\phi_{12}}$.

The latter are the two interesting cases. In the first case, both sources are assumed to be similar and uncorrelated. In the second case, the reference phase noise is much smaller than that of the source being tested and its contribution can be neglected. In both cases, it is important that the sources are really independent and uncorrelated as the system cannot detect common modulations on the two sources. Effects of injection locking due to inadequate isolation can lead to erroneous results. Nevertheless, injection locking effects can be detected if the feedback loop is opened at the loop filter input and very slow beat notes are observed by means of a strip chart recorder. Any tendency to undesired locking will show up as irregularities in the beat signal.

If these precautions are observed and if the loop gain and time constant are adjusted so that

$$(2\pi f)^2 \, T_1 \gg K_0 \tag{7.4}$$

for the range of Fourier frequencies f we are interested in, then $H_l^2 = 1$ and

$$S_{\phi_{12}} = \frac{S_{U_0}}{K_d^2}. \tag{7.5}$$

The measurement of S_{U_0} is made by means of a low frequency wave analyser. Suitable analogue wave analysers are available having a resolution bandwidth as low as 1 Hz and an operating frequency range from 2 Hz up to a few tens of kHz. Narrower bandwidths and lower operating frequency range limits are beyond the capabilities of usual commercial apparatus and not very practical to use because of the time consuming measurements required. In order to remain within the recommended scan speed limits mentioned in Section 7.2, a single analysis run from 2 to 5000 Hz with a resolution of 1 Hz takes about 20 minutes. This is the main reason for which fluctuations slower than $f \sim 1$ Hz are investigated best through analysis of time-domain data.

The low-frequency limit of $f \sim 2$ Hz gives an indication as to the choice of T_1 and K_0. For measurements on high quality crystal oscillators operating at 5 MHz, values of $T_1 \sim 10\,s$ and $K_0 \sim 1\,s^{-1}$ are useful orders of magnitude. This means that the phase-lock of this measuring loop is *very loose*, its purpose being only to keep the two input signals in exact quadrature on the long term, whereas for the fast fluctuations which have to be investigated with this method, the sources behave as if they were free running.

The upper limit of measurement frequency range is determined by the bandwidth of the d.c. amplifier. A range of 50 to 100 kHz is sufficient for most actual low-noise applications.

The calibration of the system is quite easy: in the open loop, a slow beat voltage is observed at the output of the amplifier.

Let U_{opp} be the peak-to-peak voltage of this beat note. K_d is then given by

$$K_d = \left(\frac{\mathrm{d}U_0(\phi)}{\mathrm{d}\phi}\right)_{\phi=0} \quad \text{with } U_0(\phi) = \tfrac{1}{2}U_{\mathrm{opp}}\sin\phi$$

$$K_d = \tfrac{1}{2}U_{\mathrm{opp}} \quad \text{(V/radian)}. \tag{7.6}$$

On the other hand, the power spectral density of the voltage U_0, in the closed loop, is equal to

$$S_{U_0} = U_{\mathrm{WA}\,1}^2 \quad \text{(V}^2/\text{Hz)} \tag{7.7}$$

where $U_{\mathrm{WA}\,1}$ is the RMS voltage measured for 1 Hz noise bandwidth of the wave analyser. For other analyser noise bandwidths B (in Hz), we measure an RMS voltage $U_{\mathrm{WA}\,B}$ and have to convert:

$$S_{U_0} = \frac{U_{\mathrm{WA}\,B}^2}{B^2} \tag{7.8}$$

and we obtain

$$S_{\phi_{12}} = \frac{4U_{\mathrm{WA}\,B}^2}{B^2 U_{\mathrm{opp}}^2} \quad \text{(radian}^2/\text{Hz)} \tag{7.9}$$

for a general bandwidth of B Hz and

$$S_{\phi_{12}} = \frac{4U_{\mathrm{WA}\,1}^2}{U_{\mathrm{opp}}^2} \quad \text{(radian}^2/\text{Hz)}. \tag{7.10}$$

for 1 Hz bandwidth.

As shown in Section 2.5, the result of low-level phase noise measurements can also be expressed as a sideband to total signal power ratio per one Hertz bandwidth through the relation

$$\mathscr{L}(f) = \tfrac{1}{2}S_\phi$$

valid if the total RMS phase deviation is small compared to 1 radian. For many applications, especially in telecommunications, it is customary and practical to express this power ratio in decibels. We again have to distinguish between the two cases:

Case 1. $S_{\phi_1} = S_{\phi_2}$, thus $S_{\phi_1} = \frac{1}{2}S_{\phi_{12}}$

$$\mathscr{L}(f) = 20 \log \frac{U_{WAB}}{U_{opp}} - 20 \log B \text{ (dB)}. \tag{7.11}$$

Case 2. $S_{\phi_1} \gg S_{\phi_2}$, $\quad S_{\phi_1} \approx S_{\phi_{12}}$

$$\mathscr{L}(f) = 20 \log \frac{U_{WAB}}{U_{opp}} + 3 - 20 \log B \text{ (dB)} \tag{7.12}$$

As shown in Ref. 1 for Case 1, the computation of the result is facilitated by adjusting U_{opp} (by setting the d.c. amplifier gain) to a convenient value. For a bandwidth of $B = 1$ Hz, the following relation is obtained:

$$U_{WA1} = \quad 10 \quad\quad 100 \quad\quad 1000 \quad nV/\sqrt{Hz}$$
$$\mathscr{L}(f) = -150 \quad -130 \quad -110 \quad dB$$

if we set

$$U_{opp} = \begin{array}{l} 0\cdot316 \text{ V for Case 1} \\ 0\cdot447 \text{ V for Case 2}. \end{array}$$

The order of magnitudes of the voltages shown are realistic experimental values which are typical for measurements made at 5 MHz. This illustrates the high sensitivity of the system which is limited by the phase noise of the reference source and the noise of the mixer and d.c. amplifier. The noise levels of the latter can be measured by feeding the same signal into both mixer input ports and recording the wave analyser output voltage. In the current state of the art the critical element contributing to the system noise is the d.c. amplifier input stage, especially at low frequencies (below 100 Hz) where its flicker noise becomes dominant. The mixer usually has a lower noise level than the amplifier, provided it is operated with optimum input signal level and correctly terminated. The optimum value of the load resistor depends on the mixer and the amplifier; some experimental work is therefore required to reach the optimum.

Needless to say that the measurement of such extremely low noise levels (orders of nanovolts in 1 Hz bandwidth) requires much care in the design and adjustment of the equipment. Stray field pickup due to ground loops and improper grounding must be avoided. This requires careful shielding of all leads, operating all amplifiers individually on built-in batteries and, in extreme cases of ambient noise environment, installing the whole setup in a Faraday cage.

Figure 7.3 shows a schematic diagram of a low-noise d.c. amplifier of recent

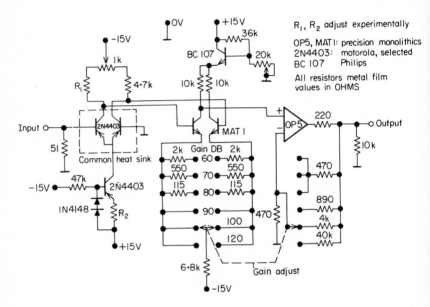

FIG. 7.3. Low-noise d.c. amplifier—schematic diagram.

design[2] having a low flicker noise level. The noise voltage level and the corresponding system noise level are shown in Fig. 7.4 for the frequency range from 2 to 5000 Hz.

The phase-noise of the reference source is in most cases higher than the system noise but it is very useful to have a very low-noise system available for its application as a sensitive differential phase-noise measurement system. This application is shown in Fig. 7.5. Here, the output of the reference signal source is split. One branch is applied directly to the mixer and the other sent through the two-port device under test. This arrangement allows the

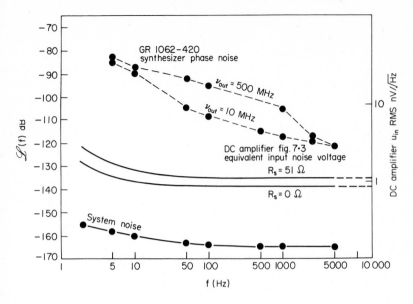

FIG. 7.4. D.c.-amplifier, PLL-system and synthesizer noise.

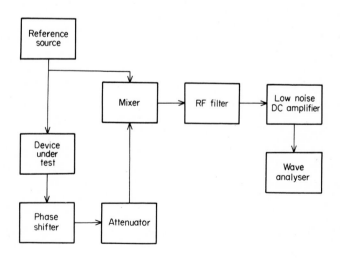

FIG. 7.5. Differential phase-noise measurement.

measurement of differential phase noise for a great variety of passive and active components, the only limitations being the frequency range of the available reference source and the useful input and output signal levels. The reference source must have low noise because of the possible signal delay through the test object. Otherwise, noise would be observed which is due to the signal source and not to the tested device owing to decorrelation of the signals.

Differential phase noise measurements are a basic requirement for all work aimed at improving the phase-noise performance of oscillators, amplifiers and various components; the method described here can therefore serve as a valuable tool for further work.

Most of the measurements and development work on very low phase noise oscillators has been done originally on 5 MHz. The PLL-method is however much more universal. If measurements are to be made on a wider frequency range but with less demanding requirements on the phase-noise level of the reference source, a good low-noise frequency synthetizer can be inserted between a 5 MHz reference VCO and the mixer. The verified performance of state-of-the-art synthetizers is indicated as a dotted line in Fig. 7.4.

Low phase–noise measurements can be made at any frequency suitable for the mixer if two similar sources are available, one of which is voltage controllable. The system parameters will have to be established for each case. The method can also be extended into the microwave range by means of appropriate signal sources and heterodyne converters, as shown in Ref. 1. In the same reference, another method of phase-noise measurement technique using a microwave resonant cavity in a bridge circuit as sensitive phase discriminator is also described.

7.4. USE OF THE PHASE-LOCKED LOOP FOR TIME-DOMAIN MEASUREMENTS

The phase-locked loop method described in the preceding Section 7.3 can also be adapted to time-domain measurements of frequency fluctuations. Two modifications of the system are required. First, the reference source must be very strongly locked to the source under test. This is obtained by shortening the loop filter time constant T_1 and increasing the loop gain K_0. We then obtain in a somewhat restricted range of Fourier frequencies:

$$f \ll \frac{K_0}{2\pi} \quad \text{and} \quad (2\pi f)^2 \, T_1 \ll K_0$$

$$|H_l(2\pi i f)|^2 \approx \left(\frac{2\pi f}{K_0}\right)^2 \tag{7.13}$$

and thus

$$S_{U_0}(f) \approx \left(\frac{2\pi f}{K_2}\right)^2 S_{\phi_{12}} = \left(\frac{2\pi \nu_0}{K_2}\right)^2 S_{y_{12}}(f). \qquad (7.14)$$

This means that the random fluctuations of the voltage U_0 are proportional to the relative frequency offset fluctuations y_{12} of the two compared signals, since their spectral densities are proportional.

Sample averages of U_0 over variable sampling times τ can be measured by the means shown in Fig. 7.6. Instead of the wave analyser, a voltage to

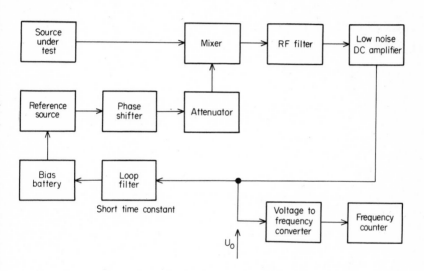

FIG. 7.6. Use of the phase-locked loop for time-domain measurements.

frequency converter is connected to the output of the d.c. amplifier and this frequency is measured with a counter in the direct or multiple period mode (see Chapter 5). It must however be kept in mind that the method is applicable only for the Fourier-frequency range for which (7.13) is valid, i.e. very short time averages τ must be approached with caution. The advantage of the method is its high sensitivity and the ease of conversion of an existing phase-noise measurement system. In many cases, a restricted range of Fourier frequencies exists, where both modes of operation can be applied (sequentially, not simultaneously) and this enables the validity of the measurements to be checked by using the relations (2.19) and (2.35), etc. (see Chapter 2) to compute $Sy(f)$ from $\sigma_y^2(\tau)$ and vice versa. If the results are consistent, much confidence is gained about the validity of the system. Otherwise, sources of unwanted interference, etc., will have to be considered. Problems of this kind will usually show up in the phase-noise spectral density measurements, e.g. peaks at the power line frequency and its harmonics.

REFERENCES

1. J. H. Shoaf, D. Halford and A. S. Risley. "Frequency Stability Specification and Measurement: High Frequency and Microwave Signals", NBS Technical Note 632. US Govt Printing Office, Washington, DC, January 1973. (See also Refs. 14 and 15 of Chapter 2).
2. K. Hilty and J. Ph. Mellana. Swiss PTT R & D Division, 3000 Bern 29, Switzerland. Private communication (to be published).

8

Radio Signal
Comparison Methods

8.1. INTRODUCTION

Time and frequency comparison methods over some distance are closely
related to the techniques of time and frequency dissemination (TFD).
Dissemination and comparison by means of radio communications existed
since the early beginnings of radio around the turn of the century, replacing
most of the older techniques used since the invention of wire telegraphy.
The dynamic nature of time-keeping, the need for synchronization and the
fact that no clock is perfect requires dissemination of coordinated timing
information. The modern industrial society could not function without it.

The variety of systems and methods is as great as that of the numerous
users' requirements which span a wide range of precision, availability and
reliability. No single universal system has yet been conceived which would
satisfy all potential users to the same extent and it is doubtful whether this
will ever be possible.

Basic methods of TFD can be classified into three main categories of
unequal importance, namely:

Physical transport of operating clocks.
Wire communications.
Radio communications.

It is somewhat like ordinary information transmission, except that time is a
special kind of information, i.e. physical transport does not mean sending
time in a letter or a parcel but carrying a running clock synchronized at a

179

reference point and transported so carefully that it keeps its time scale during the trip.

Clock transport is actually still one of the most accurate and reliable methods of transferring time (in the meaning of date) from one place to another. Conceptually, the technique is very simple. The clock is synchronized at the reference station (observatory or standards laboratory) and carried to the desired places. Synchronization and comparison at the various test sites are made by means of a time interval counter. Usually a round trip is made back to the reference station where another comparison yields the "time closure" of the trip, i.e. the clock error accumulated during the trip. This serves as a measure of confidence about the errors due to the clock during the operation. Environmental effects on the clock's rate must of course be known and kept as small as possible. These may be due to temperature, magnetic field, acceleration, atmospheric pressure (air transport), etc. The performance of modern commercial caesium clocks is now so high that even relativistic effects may be taken into account, the effect of the gravitational red-shift near the surface of the Earth being about 1 part in 10^{13} per kilometer change in altitude. The stability of a caesium standard can be better than 1 part in 10^{12} (timing error less than 86·4 ns per day) during a normal trip.

Clock transport is not a cheap way of communicating time from place to place but it is not the most expensive one either. Many trips can be paid for the cost of setting up a satellite receiving station for the purpose of micro-second timing and then requesting a clock trip anyway because nobody can guarantee that the transmission delay computations are really correct. Besides its basic simplicity, the method has its special excitements, e.g. running for the nearest power line outlet because the batteries are running low (time lost = go home and get scolded) or trying to convince a customs officer that an atomic clock is not an A-bomb and by far less radioactive than his (the officer's) wrist watch. The clock transport method is widely used for initial calibration and periodic verification of important extended timing systems, e.g. radio navigation and satellite tracking networks, where micro-second timing accuracy is a vital requirement.

Wire communications as a means of time and frequency dissemination is mostly used for low precision public time services. In this category we find the well-known time announcements offered by telephone companies which are accurate to a few milliseconds and also the power line network driving synchronous clocks to within a few seconds. The potential accuracy of wire communication links is, however, not fully exploited, except in some special applications within long distance telephone networks. In carrier telephony systems, standard frequencies are distributed to control the various carrier frequency generators to a few parts in 10^9. It is quite probable that new developments will take place in this field with the future developments of large switched digital communications networks.

Radio signals as a means of TFD and comparison of remote clocks and frequency standards represent the largest and generally most important category of comparison methods. This is mainly due to the fact that for the individual user these methods represent an optimum from the standpoint of flexibility and cost-effectiveness.

The variety of systems, services and methods is so great that in order to keep this chapter within a reasonable size, the remaining sections will be restricted to a condensed review of what the author feels are the currently most important methods. For those readers wishing more complete and detailed information, reference is made to the very complete review written by Byron E. Blair in Ref. 1 (80 pages, 281 cited references) and to the documents of the Study Group 7 of the CCIR whose official terms of reference are devoted to standard frequency and time signals.

Radio signal TFD systems and comparison methods can be categorized in different ways, e.g. either according to the users' requirements or to system specifications including cost of implementation and operation. Distinction can also be made between systems which are exclusively dedicated to TFD and systems which serve a different chief purpose but rely on precise timing. In this latter class we find several precision navigation systems as LORAN-C and Omega, television networks and some satellite communication systems.

In this chapter, the principal characteristics of the following systems or methods are reviewed:

HF-radio systems
CW–LF radio systems
LORAN-C
VLF systems
TV-timing
Satellite systems.

8.2. HF STANDARD FREQUENCY AND TIME SIGNALS

In the HF band of radio waves extending from 3 to 30 MHz (decametric waves, wavelength 100 to 10 metres), long distance propagation mainly takes place by reflection from the ionosphere, i.e. "sky-wave" propagation in single or multiple hops between the surface of the Earth and the ionosphere.

Ground-wave propagation is noticeable only over short distances. The field strength of the ground-wave diminishes rapidly with increasing distance from the transmitting antenna.

The characteristics of HF-radio propagation vary strongly as a function of frequency and of time as is well known to every listener on short-wave broadcasting bands and to professional and amateur radio operators. On the other hand, world-wide coverage is possible with moderate transmitter power (usually less than 20 kW) and simple receiving equipment.

In the currently valid International Radio Regulations the following frequency channels have been allocated by the Administrative Radio Conference (Geneva 1959) to standard frequency and time signal services in the MF and HF bands*:

2·5 MHz $\pm$5 kHz (Note 1)
5   MHz $\pm$5 kHz
10   MHz $\pm$5 kHz
15   MHz $\pm$10 kHz
20   MHz $\pm$10 kHz
25   MHz $\pm$10 kHz

Note 1: In the ITU Region 1 the channel width is reduced to $\pm$2 kHz. Region 1 includes Europe, Africa and the Asian territories of Turkey, USSR and the Mongolian Peoples' Republic.

Additional standard frequencies and time signals are emitted in other frequency bands. Detailed information on standard time and frequency emissions is published at regular intervals:

Every 4 years by the CCIR in its Plenary Assembly Proceedings, latest issue Geneva 1974

every year by the BIH in its annual report.

Appendix 8.2 reproduces Part C of the BIH Annual Report for 1974† which contains the text of the CCIR Recommendation 460 concerning Standard Frequency and Time Signal Emissions and detailed information on the transmitters operating in 1974. Many of these transmitters have been in operation for several years but some changes occur from time to time. The reader is therefore advised to consult the most recent issue of the publications mentioned above.

* For the official CCIR designations of radio bands, see Appendix 8.1.
† The permission of Dr B. Guinot, Director of the BIH and of the Director, CCIR to reproduce this text is hereby gratefully acknowledged.

In the HF band, the obtainable precision of time and frequency comparisons is almost entirely limited by propagation phenomena due to the unstable nature of the ionosphere. The phase of the received carrier signal fluctuates because of the variations in path length and propagation velocity. These fluctuations limit the highest achievable precision to about $\pm 1 \times 10^{-7}$ for frequency comparisons and to about 500 to 1000 μs for the reception of timing pulses.

These values appear quite modest to specialists in instrumentation used to and impressed by the most recent developments; however, this limit of precision satisfies the needs of a large majority of users, especially in maritime navigation. As shown on the list in Appendix 8.1, there are many transmissions on a large number of frequencies throughout the HF band and the transmitter locations are distributed over both hemispheres of the Earth. A radio operator experienced in HF communications can almost at any time of the day find a useful signal, provided he chooses the right frequency according to the propagation conditions. In some regions, especially in and around Europe, the proliferation of transmitters is such that mutual interference has become a major problem. The CCIR in its Recommendation 374-2 (1974) has therefore recommended that the administrations responsible for the emissions consider alternative methods of disseminating standard frequencies and time signals before adding new emissions in bands 6 (300–3000 kHz) and 7 (3–30 MHz).

Figure 8.1(a) shows a block diagram of a simple HF-radio time comparison system including a HF communications receiver, an oscilloscope and a digital delay generator connected to the local clock. The delay generator is a preset counter (see Chapter 5) generating an adjustable 1 p.p.s. output pulse for triggering the oscilloscope time base. The receiver output is connected to the y-axis input of the oscilloscope. A loudspeaker or earphone is required for direct monitoring and identification of the received signal.

The usual timing information carried on the transmitter signal consists of short audio frequency bursts, e.g. 5 cycles of a 1000 Hz wave in the case of the WWV transmitter system (see Appendix 8.2) which can be heard as "ticks" in the audio signal. The time comparison is very simple. Once the ticks have been clearly heard above the noise, the oscilloscope sweep speed and the delay are adjusted until the signal appears on the screen. It is then shifted to the beginning of the trace by means of fine delay adjustment and the difference can be read from the positions of the delay generator switches. This gives the time of arrival of the signal *at the receiver output* with respect to the pulse of the local clock, to the nearest second. Minute identification can be checked by waiting for the minute markers.

For use with more elaborate systems, some services transmit a full digital time code (see Section 4.5).

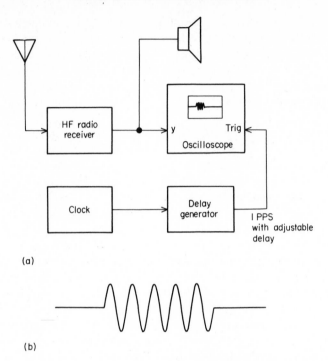

(a)

(b)

FIG. 8.1. HF time signal reception.

In order to know the time of the local clock with reference to UTC, the propagation delay has to be taken into account. The time services indicate the error limits of their signal with reference to UTC. The computation of the path delay due to the distance separating the transmitter from the receiver must be done by the user. For the precision of ± 1 ms, the computation of the path delay is relatively simple. Great circle distances between points defined by their coordinates in longitude and latitude can be computed assuming a spherical Earth according to the formula:

$$\sin^2 \frac{\delta}{2} = \cos \beta_A \cos \beta_B \sin^2 \frac{\lambda}{2} + \sin^2 \frac{\beta}{2} \tag{8.1}$$

where:
Point A has latitude β_A and longitude λ_A;
Point B has latitude β_B and longitude λ_B.
The angles are expressed in degrees, with the signs:

Latitude is North positive and South negative;
Longitude is East positive and West negative.

β and λ are the absolute latitude and longitude differences between points A and B:

$$\beta = |\beta_A - \beta_B|,$$
$$\lambda = |\lambda_A - \lambda_B|.$$

The result is the angle δ encompassing the great circle between A and B. If we express δ in minutes of arc, we get the great circle distance D in nautical miles and in kilometres, we have

$$D = 1\cdot852\delta \,(\text{km})$$

$$\text{since } 1\text{n.m.} = 1\cdot852 \,\text{km}$$

again with δ expressed in minutes of arc.

The effective ground wave propagation speed is slightly below the vacuum speed of light. A value of 299·773 km/s is usually assumed.[1] Ionospheric propagation is accounted for by corrections which do not exceed 1 ms for distances above 1000 km. Figure 8.2 taken from Ref. 2. and reproduced in Ref. 1 gives the path delay for single hop transmissions. In case of multiple hop propagation, the distance must be divided by the number of hops, the delay taken for the individual hops and then again multiplied. The determination of the number of hops is not always easy and simultaneous presence of several modes will result in multiple tick reception or time jitter between relatively constant positions.

The following rules[1] are helpful in ensuring optimum reception and measurement precision:

1. The measurements should be made at the same time every day when the radio path is either in full daylight or darkness.
2. No measurements should be made during ionospheric disturbances.
3. The highest useful radio frequency should be used.
4. Paths passing through the polar auroral zones are to be avoided.
5. A good receiver and a directional antenna should be used.

If a HF-radio signal is to be used as a reference for frequency measurements, the signal can be injected at a low level into the antenna circuit of the receiver and adjusted until a zero beat frequency is obtained in the audio output. However, this can be safely done only during periods when the carrier is not modulated. Many signals carry audio frequency modulation during part of the transmission time. There is then a risk of error due to "zero beating" on a sideband instead of the carrier. As stated before, the precision of frequency measurement is limited to about $\pm 1 \times 10^{-7}$ on sky-wave signals. Higher

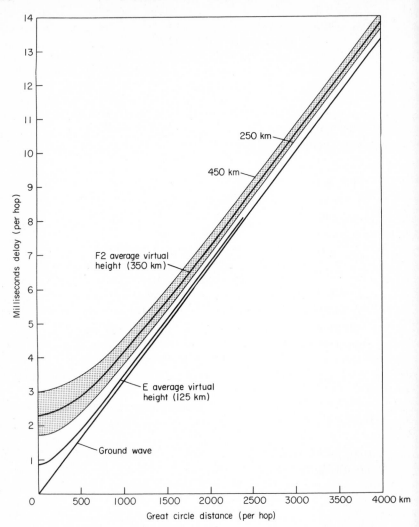

FIG. 8.2. Path delay for single hop transmissions.

precision is possible in the close vicinity of the transmitter but this concerns only a small minority of users.

In restricted regions where the signals are available, far better and more reliable results are obtained by using other frequency bands and methods as described in the subsequent sections.

8.3. LOW FREQUENCY CW SYSTEMS

The low frequency (LF) band (Band 5, 30–300 kHz) is very well suited for regional standard frequency and time dissemination. Somewhat higher transmitter powers (25–100 kW) and more expensive antennas are required than for HF transmitters.

Some of the stations now in use for TFD are former radio-telegraphy transmitters reconverted to this service. The propagation characteristics of LF waves have been known for their stability and reliability since the early times of wireless radio communications. The widespread use of the LF band for TFD is relatively recent, although some services such as MSF on 60 kHz have operated for 25 years. Its introduction has been developed in parallel to the evolution of some radio-navigation systems based on time or time difference measurements. Obviously such navigation systems also must rely on the stability of the wave propagation. The development of TFD and navigation systems has therefore not taken place independently but with strong mutual interaction. One of the best time comparison methods currently in use relies on the LORAN-C system which operates with a 100 kHz carrier frequency. This method will be reviewed in Section 8.4. In the present section we shall restrict the review, as the title says, to CW systems. More specifically, by CW, we do not mean a purely continuous wave transmission, but signals where the carrier is modulated by slow amplitude or frequency shift "keying" (as in telegraphy) so that the signal spectrum occupies a narrow channel only a few hundred hertz wide, in compliance with the regulations for radio telegraphy operations (class A1 or F1 emissions).

The relatively recent introduction of TFD in the LF-band is also reflected in the fact that no exclusive specific TFD allocations actually exist for frequencies in this band. The dedicated TFD services are operating now either on an experimental or on a permanently tolerated basis which is justified by their usefulness. The transmitters operating in the LF band are listed in the Appendix 8.2. Most services transmit seconds pulses with minute markers and additional timing information such as DUT1 and some others have included a full BCD time code in their emissions.

Depending on the application the user requires, the LF signals offer the two possibilities of either frequency or time information as with HF signals but on a higher level of precision. The range of coverage is limited to about 1000 km for daytime ground-wave propagation and to about 2500 km for night-time sky-wave propagation. Phase fluctuations due to variations of ionospheric propagation are much less marked than on HF signals. Their extent depends on the distance separating the receiving site from the transmitter and on the exposure of the path to the radiation from the Sun. These

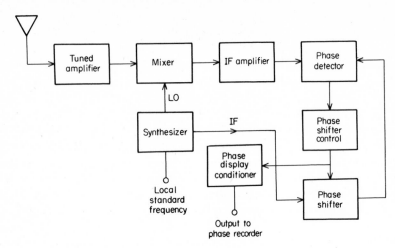

(a)   LF or VLF phase comparison receiver

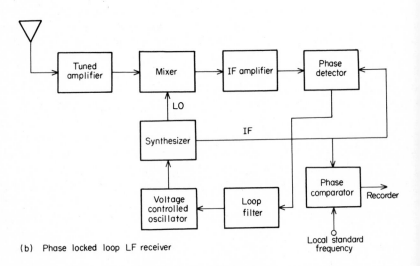

(b)   Phase locked loop LF receiver

FIG. 8.3. Phase tracking receivers.

effects are to be considered if use is made of the carrier signals as a standard frequency source. The generally accepted method for this use is phase comparison by means of a phase-tracking receiver, and it is valid not only for CW and pulsed LF systems but also for VLF comparisons discussed in Section 8.5. Figure 8.3 shows two basic types of phase-tracking systems for LF and VLF phase or phase-time comparison. The principle shown in Fig. 8.3(a) is used in some commercial phase comparison receivers. The signal picked up by the antenna is amplified in a bandpass amplifier and then converted to a constant intermediate frequency. The desired transmitter frequency is selected by changing the local oscillator frequency which is generated by means of a synthesizer driven by the local standard frequency. This is necessary in order to preserve the phase relationship between the received signals and the local signals. A free running local oscillator would obviously not do this. The intermediate frequency is preferably chosen to have a round-figure value, e.g. 10 kHz, in order to facilitate the interpretation of the results as shown below. The synthesizer also generates an intermediate frequency signal which is applied to a phase detector through an adjustable phase shifter. The output d.c. voltage of the phase detector controls the phase shifter and shifts the phase of the local i.f. signal to achieve synchronism between the received i.f. and the local i.f. signals.

The position of the phase shifter control thus follows or "tracks" the phase of the input signal with respect to the phase of the local signal and represents therefore the phase (degrees) or phase-time (microseconds) difference between the signals. In older systems, mechanical rotating phase shifters (synchro resolvers) driven by a d.c. motor were used. A d.c. voltage proportional to the angular position of the phase shifter was obtained by means of a potentiometer coupled to the shaft of the resolver. In recent designs we find electronic phase shifters using digital logic circuits, e.g. frequency dividers, the output phase of which can be shifted in steps by either doubling or suppressing some pulses in the input signal. If the output phase-time of a 10 kHz signal is to be shifted in steps of 0·1 μs, an input pulse repetition frequency of 10 MHz and a division ratio of 1:1000 are required. The d.c. output voltage of the phase detector is applied to a threshold sensor which generates "advance" or "retard" shift pulses if the voltage exceeds a minimum positive or negative value. The rate of these shift pulses determines the gain and the time constant of the tracking loop. An analogue d.c. voltage for displaying the position of the digital phase shifter is generated by means of a reversible counter registering the shift pulses and a D/A converter which is connected to the counter.

This short description of a small part of a fully electronic phase-tracking receiver shows that such systems can be quite elaborate, sophisticated and relatively expensive. Depending on the local and receiving frequencies and on the structure of the synthesizer, such a receiver can have a multitude of equilibrium positions of the phase shifter. If the local reference frequency is interrupted, the tracking system does not therefore necessarily return to the previous value.

Becker and Kramer[3] have developed a system which circumvents this disadvantage. The desired LF or VLF signal is filtered and amplified directly without frequency conversion and the phase detector is given by sampling pulses at a rate of only 100 Hz, a rate that allows tracking of all signals having frequencies which are exact multiples of 100 Hz, a condition satisfied by most currently operating transmitters.

Another possibility of phase comparison receivers is shown in Fig. 8.3(b). Here, a local oscillator is locked to the received carrier phase using the principle of the PLL synthesizer, as described in Section 3.5 (Fig. 3.27). If a good signal is available at the antenna, the local oscillator can be of moderate stability, e.g. a simple crystal oscillator without temperature control. If the local oscillator is locked, its output phase remains stable with respect to the received signal phase. The comparison with the local standard frequency is then possible by means of a phase-time comparator of the type described in Chapter 6 (Fig. 6.3).

Phase-lock receivers of this type can be of a relatively simple design if they are tailored to a particular transmitter frequency. They also offer the advantage of generating a stable output signal frequency locked to the received carrier. The disadvantage is again the ambiguity between several possible equilibrium phases of the local oscillator with reference to the received carrier. The possibility of locking on another phase relationship after an interruption of the received signals constitutes a problem of reliability, especially if continuous phase recordings are to be made over longer periods of time. Where only occasional measurements are to be made over a limited time and the highest reliability is not required, phase-lock receivers constitute an inexpensive alternative to the more elaborate designs of Fig. 8.3(a).

Figure 8.4 shows the time dependence of the phase of a LF signal over a distance of a few hundred kilometres where both ground-wave and sky-wave signals are present at the receiving site.

There is a marked difference between day and night with the shift at sunrise being faster than at sunset. Furthermore, the stability of the phase is best during daytime when the entire path is in sunlight. During night time, the fluctuations due to ionospheric instabilities are stronger, especially at distances where the skywave amplitude dominates that of the ground-wave. During periods of increased solar activity, some phase instabilities can also be observed during daytime. During stable reception conditions, the day-to-day repeatability of phase measurements seldom exceeds one microsecond. This allows as a general rule, a precision of frequency comparisons of about 1×10^{-11} in 24 hours. At short distances from the transmitter slightly better precision can be obtained. However, in CW transmissions, it is not possible to separate the reception of the ground-wave from that of the sky-wave as it is done in the reception of LORAN-C pulses (see next Section 8.4).

The bandwidth available for radio-telegraphy also allows the transmission of timing information. The good propagation characteristics of the LF band allows a significant improvement in the precision of time signal reception compared to that possible on HF time signals. The possible precision has been

studied in detail by Andrews and De Prins[4] on the time signals of WWVB*
and HBG* at distances up to about 500 km from the transmitter. The reference time on the signals used is marked by a carrier interruption. The total
decay time due to the resonant behaviour of the transmitter output and
antenna circuit is of the order of 2 ms. Careful measurement of the signal

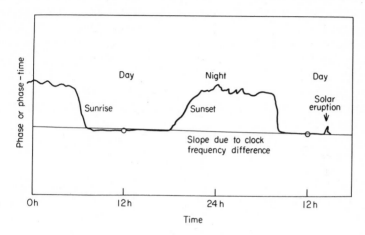

FIG. 8.4. LF or VLF phase record (example).

amplitude and waveform by averaging a large number of pulses allows the
determination of a time reference point with a probable error of about
$\pm 40\ \mu s$. The optimum level for the time measurement was found to be
around 85 per cent of the carrier amplitude. Even with very simple receivers
and less precautions, it is possible to achieve a repeatability of time measurements of about $\pm 200\ \mu s$ on LF time signals under good conditions of
reception. The main problem encountered in the use of LF timing signals is
the high level of man-made noise present in most industrial and laboratory
environments. As the effects of this type of noise are of a random nature, they
can be reduced more easily for carrier phase measurements than for the
time signal reception. On account of the impossibility of predicting the man-
made noise level at potential users' locations, it is difficult to indicate the
useful range of a LF time signal transmission. The author has seen people
having difficulties in receiving HBG in downtown Geneva, only about 20 km
away from the transmitter whilst he observed clean signals, from the same
transmitter, at a location near Helsinki, Finland.

* For characteristics, see Appendix 8.2.

8.4. LORAN-C

Also operating in the LF band, the LORAN-C (LOng RAnge Navigation) navigation system is of special importance as a means of precise time and frequency comparison. This is mainly due to the fact that the LORAN-C system uses fast pulse transmission and not quasi-CW signals. The carrier frequency is 100 kHz and the signal spectrum contains significant components in the band from 90 to 110 kHz.

The rather intricate detailed characteristics of the signals, together with synchronous and coherent reception techniques result in a good resistance to interference—a welcome property since in Europe several CW transmitters are operating in the 90–110 kHz band, some with much higher power than that of the LORAN-C stations.

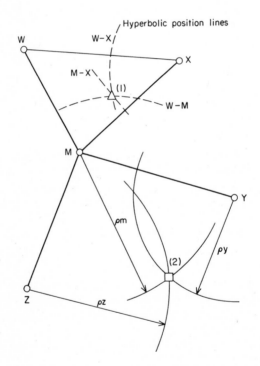

(1) △ Vessel working with hyperbolic navigation

(2) □ Vessel working with RHO – RHO navigation

FIG. 8.5. LORAN-C chain schematic.

As we shall see, the short risetime of the pulses allows a time-domain separation between the ground-wave and the remaining sky-wave components, since the ground-wave always travels on the shortest path and therefore arrives first. The possibility of making phase-time measurements of the ground-wave signal at distances exceeding 800 km from the transmitter over a land path (longer range over sea water paths) result in a typical precision of ± 0.1 μs, i.e. the errors due to propagation effects are roughly ten times smaller than those observed on LF-CW transmissions.

Furthermore, the LORAN-C signals are shaped in a way that reliable detection of a particular zero-crossing is possible in principle. In practice, there are sometimes certain difficulties in the initial detection of a particular cycle but once established for a given receiving equipment, including the antenna, it is usually always possible to retrieve the same cycle in the signal even after some interruption.

These properties make the LORAN-C system very interesting not only as a means of frequency comparison via phase-time measurement, but also for dissemination of precise timing information.

The main disadvantage of this system is that its coverage is limited to the regions where LORAN-C chains are in operation, namely Europe, North Atlantic, Caribbean, North, Central and West Pacific. There are no systems operating in the southern Hemisphere. An extension of this system to the US West Coast is currently under way.

The following short description of LORAN-C techniques is based on the papers by Shapiro,[5] Potts and Wieder[6] and Blair[1] supplemented by a few years of practical experience in using one LORAN-C station to monitor the performance of a commercial caesium frequency standard (for some results, see Example 2 in Section 4.4).

A LORAN network or chain consists of a master (M) station and two to four slave stations (X, Y, Z, W), as shown schematically in Fig. 8.5.

The network is synchronized from the master station which, at the beginning of each group repetition interval (GRI), transmits a group of pulses. Then, in turn, each slave transmits its pulse group with increasing delay so that within the operating region of the chain, the various groups are always received in the same sequence and do not overlap in time. The emission delay of each slave is the sum of: (a) the propagation delay from master to slave, and (b) an additional "coding delay" generated at the slave station. At an arbitrary receiving site the signals from the various transmitters suffer an additional propagation delay, which when measured, yields the basic information for the navigator's position fix.

There are two basic methods of navigation using such a synchronized network, namely hyperbolic navigation and range–range (rho–rho) navigation.

Hyperbolic navigation is based on the measurement of propagation delay differences, using the well-known geometrical principle that all points having a constant difference of distance to the two baseline end points (i.e. the transmitter sites) define a hyperbola of which the baseline end points are the foci. Using the corresponding time (and thus distance) differences from three stations, three hyperbolic lines of position are determined, which, if everything is right, intersect at the observer's position. Rho–rho navigation requires an on-board clock synchronized to the transmitter system, allowing to measure the real distance to the transmitters. Hereby, the position is obtained as an intersection of circles drawn around the stations instead of the hyperbolae.

With both methods, distance measurements are obtained from time difference measurements using the known speed of wave propagation. It is thus easy to understand why accurate synchronization is a primary requirement in such navigation systems. One microsecond of timing error leads to a distance error of 300 m (1000 ft) and the resulting position error can be much larger in some geometrical configurations where lines of position intersect at small angles.

In order to ensure accurate master to slave synchronization, the distances between the stations are kept within the range of groundwave propagation whenever possible.

Figure 8.6 shows the pulse pattern of a typical LORAN-C chain. The master station transmits a group of eight pulses spaced at 1 ms, followed by a ninth pulse 2 ms after the pulse No. 8. The slaves transmit only the group of 8,

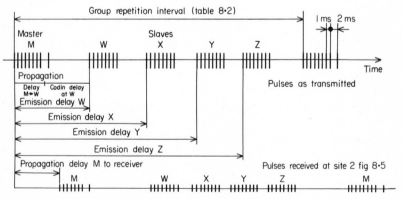

Note: In this drawing the time axis is compressed between the groups of pulses. The delays are not exact but exaggerated to illustrate the principle.

FIG. 8.6. LORAN-C pulse patterns.

TABLE 8.1. LORAN-C transmitter data (Ref. 1 and CCIR)

Chain	Rate		Station	Latitude	Longitude	Emission delay (μs)	Peak radiated power (kW)
US East Coast	SS7	M	Carolina Beach N.C.	34°03·8' N	77°54·8' W	....	1000
		W	Jupiter Fla.	27°02·0' N	80°06·9' W	13 695·48	400
		X	Cape Race NF	46°46·5' N	53°10·5' W	39 389·56	2000
		Y	Nantucket Is. MA	41°15·2' N	69°58·6' W	52 541·27	400
		Z	Dana IN	39°51·1' N	87°29·2' W	68 560·68	400
Mediterranean Sea	SL1	M	Simeri Crichi, Italy	38°52·3' N	16°43·1' E	....	250
		X	Lampedusa, Italy	35°31·3' N	12°31·5' E	12 757·12	400
		Y	Targabarun, Turkey	40°58·3' N	27°52·0' E	32 273·28	250
		Z	Estartit, Spain	42°03·6' N	3°12·3' E	50 999·68	250
Norwegian Sea	SL3	M	Ejde, Faroe Is.	62°18·0' N	7°04·5' W	....	400
		W	Sylt, Germany	54°48·5' N	8°17·6' E	30 065·69	300
		X	Boe, Norway	68°38·1' N	14°27·1' E	15 048·16	200
		Y	Sandur, Iceland	64°54·4' N	23°55·3' W	48 944·47	3000
		Z	Jan Mayen, Norway	70°54·9' N	8°44·0' W	63 216·20	200
North Atlantic	SL7	M	Angissoq. Greenland	59°59·3' N	45°10·4' W	....	500
		W	Sandur, Iceland	64°54·4' N	23°55·3' W	15 068·10	3000
		X	Ejde, Faroe Is.	62°18·0' N	7°04·5' W	27 803·10	400
		Z	Cape Race NF	46°46·5' N	53°10·5' W	48 212·80	2000
North Pacific	SH7	M	St. Paul, Pribiloff Is.	—	—	....	400
		X	Attu, Alaska	—	—	14 875·30	400
		Y	Port Clarence AK	—	—	31 069·07	1800
		Z	Sitkinak AK			42 284·39	400
Central Pacific	S1	M	Johnston Is.	16°44·7' N	169°30·5' W	....	300
		X	Upolo Pt. HI	20°14·8' N	155°53·1' W	15 972·44	300
		Y	Kure, Midway Is.	28°23·7' N	178°17·5' W	34 253·02	300
Northwest Pacific	SS3	M	Iwo Jima, Bonin Is.	24°48·1' N	141°19·5' E	....	4000
		W	Marcus Is.	24°17·1' N	153°58·9' E	15 283·94	4000
		X	Hokkaido, Japan	42°44·6' N	141°43·2' E	36 684·70	400
		Y	Gesashi, Okinawa	26°36·4' N	128°08·9' E	59 463·34	400
		Z	Yap, Caroline Is.	9°32·8' N	138°09·9' E	80 746·78	4000

each station according to its allocated emission delay. Table 8.1 gives a list of the LORAN-C chains operating in 1974. Each chain is identified by its particular group repetition rate, designed by the alphanumeric code in column 2, the meaning of which is given in Table 8.2.

TABLE 8.2. LORAN-C group repetition intervals (GRI) in μs

Basic	S	SH	SL	SS
Specific				
0	50 000	60 000	80 000	100 000
1	49 900	59 900	79 900	99 900
2	49 800	59 800	79 800	99 800
3	49 700	59 700	79 700	99 700
4	49 600	59 600	79 600	99 600
5	49 500	59 500	79 500	99 500
6	49 400	59 400	79 400	99 400
7	49 300	59 300	79 300	99 300

The No. 9 Master identifier pulse is also used as a status indicator of the chain. If something is not operating correctly, this pulse is periodically blanked to warn the user. The shape of an individual pulse is shown in Fig. 8.7 as ideally generated at the transmitter. The nominal waveform is described by the function:

$$f(t) = \pm (kt)^2 \, e^{-2}(kt - 1) \sin \omega_0 t \qquad (8.2)$$

with

$$k = \frac{1}{72 \cdot 5 \times 10^{-6}} \quad \text{and} \quad \omega_0 = 2\pi \times 10^5 \, \text{s}^{-1}$$

The + and − signs are used within the pulse sequence as a phase coding which serves to reduce the sensitivity of the system to coherent signal interference, especially from long-delay multihop sky-wave signals.[5]

Table 8.3 shows the two phase code sequences for the master and slave stations transmitted on alternate group repetition intervals.[5, 6]

In most cases, depending on distance and ionospheric propagation conditions, the sky-wave signal arrives after the ground-wave signal with a delay of more than 30 μs. The first few cycles of the received waveform are therefore due to the ground-wave signal and virtually unperturbed by ionospheric effects.

Phase-time measurements made on the early zero crossings of LORAN-C

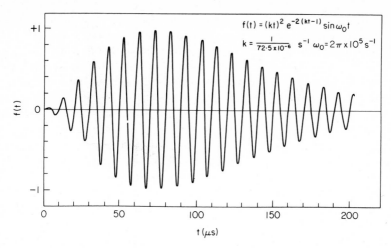

$$f(t) = (kt)^2\, e^{-2(kt-1)} \sin\omega_0 t$$

$$k = \frac{1}{72\cdot5 \times 10^{-6}}\ \mathrm{s^{-1}}\quad \omega_0 = 2\pi \times 10^5\,\mathrm{s^{-1}}$$

FIG. 8.7. Theoretical shape of LORAN-C pulse.

pulses do not show the diurnal shift between day and night transmission (see Fig. 8.4) observed on CW-LF and VLF phase recordings.

On the other hand, the techniques required to separate properly the desired ground-wave component from the rest of the signal are more involved than those sufficient for ordinary LF or VLF signals. A LORAN-C timing receiver has to perform two tasks if the advantages offered by the capabilities of the system are to be utilized:

(a) it must track the phase of the ground-wave component in the received signal; and

(b) it must enable cycles to be identified so that an unambiguous time reference is obtained, e.g. the third positive zero crossing in the signal of Fig. 8.7 (at $t = 30\ \mu s$).

The first task is relatively easy to perform by sampling the signal with a synchronous gate and coherent detection of the sample. Unambiguous cycle

TABLE 8.3. LORAN-C phase code (Ref. 6)

	First GRI	Alternate GRI
Master	+ + − − + − + −	+ + + + + − − +
Slave	+ − − + + + +	+ − + − + + − −

Note: These signs are to be applied to Eq. (8.2) in order to obtain the transmitted waveform of the pulses in the group.

H

identification is far more difficult and not always possible to the extent that portable clock calibration becomes unnecessary. However, in a system which has been calibrated once, it is possible to find the same cycle again with a high degree of confidence. The main problem is of course the fact that the ideal waveform shown in Fig. 8.7 is not necessarily preserved until it attains the detecting system. First, the propagation occurring in the medium limited by the surface of the Earth (seawater, land with various conductivities and terrain configurations, see Ref. 6) and the D-layer of the ionosphere (average height about 80 km, see Ref. 5) allows ground-wave and sky-wave propagation (single hop and also often double hop), the latter depending on the variable ionospheric conditions (night, day, solar activity).

The tail of a long delayed sky-wave component can interfere with the beginning of the following pulse. As mentioned above, this can be compensated by means of the phase alternance coding.

The most serious distortions of the waveform occur in improperly designed antenna and receiving systems. The attempt to eliminate interfering signals by means of resonant filters must be approached with caution. The delay introduced by a filter can be calibrated by means of a signal simulator. But if the filter introduces group delay distortion, the shape of the waveform is changed, especially at the beginning of the rising waveform ("precursor" signals as mentioned in Ref. 5). Proper cycle identification then becomes illusory. Figure 8.8 shows some examples of signals received in Bern, Switzerland, of the SL3-W transmitter in Sylt (see Table 8.1). In (a), the receiver has a bandwidth of 30 kHz centred at 100 kHz without any additional means of filtering. The high level of interference has been analysed at the antenna output using a spectrum analyser and the following carriers have been found (Ref. 7):

Frequency	Signal level*	Note
f_1: 75 kHz	−95 dBm	HBG
f_2: 100 kHz	−111 dBm	LORAN-C
f_3: 110 kHz	−92 dBm	
f_4: 118 kHz	−101 dBm	telegraph
f_5: 122 kHz	−100 dBm	transmitters
f_6: 132 kHz	−88 dBm	

In (b), the same signal is displayed but with a bandpass filter which is inserted in the antenna line. The filter has a passband flat to ± 0.5 dB between 90 and 110 kHz but an attenuation of 71 dB at 75 and 135 kHz respectively. The LORAN-C pulse shown in (c) is not distorted but we see the superposition

---

* At loop antenna 50 Ohm output terminal.

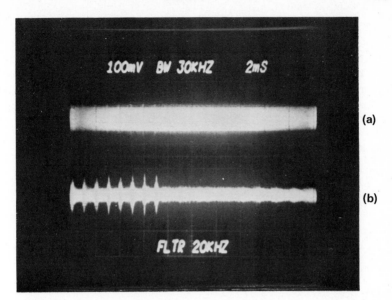

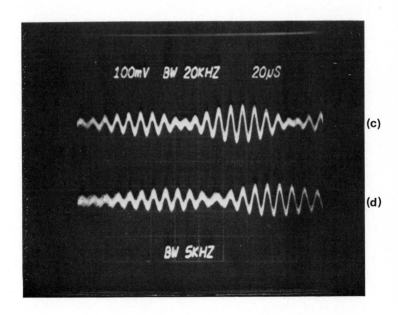

Fig. 8.8. LORAN-C signal waveforms.

of ground-wave and sky-wave components. In (d), the same pulses are filtered by a narrowband filter of a few kHz width and the shape has changed.

Figure 8.9 shows a simplified block diagram of a LORAN-C timing receiver system. A shielded loop antenna is used, giving some reduction of interference from stations located at about 90° away from the desired direction. A bandpass amplifier with automatic gain control is used to bring the

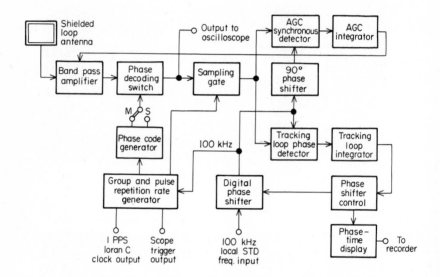

FIG. 8.9. LORAN-C tracking receiver (simplified).

signal level up to the value required by the detector and to hold it constant. The inverting electronic switch serves to decode the phase-code on the signal, as mentioned above. The decoded signal is available on an output for monitoring on an oscilloscope. At this stage, we still can observe the whole pulse or groups of pulses depending on how the scope is triggered: either by one pulse per GRI and slow sweep (50 to 200 ms/cm) or trigger bursts with the pulse repetition frequency to see superimposed individual decoded pulses.

The operation following next is to sample the first few cycles of the rising pulse waveform. This is done by the sample gate which is driven by bursts of 8 pulses, spaced at 1 ms, the burst rate being equal to the group repetition rate.

These samples are detected in two detectors, one in phase for detection of the zero crossing and one in quadrature for detection of the amplitude of the ground-wave component, in order to drive the automatic gain control. The output voltage of the zero-crossing detector is integrated and used to drive an electronic phase shifter which adjusts the phase of the 100 kHz time base frequency derived from the local frequency standard. The position of the phase shifter is displayed and an output voltage generated for the purpose of phase-time recording.

Pulse and group repetition rates, phase-code switching pulses, etc., are all generated from the phase-shifted signal and remain thus in synchronism with the received signal as long as the tracking loop is locked.

The techniques of cycle selection are based on the precise definition of the rising waveform. Under very favourable reception conditions, visual monitoring on the oscilloscope can be used but such conditions are the exception rather than the rule.[6] The use of a sampling detector to measure the respective amplitude of the first few positive and negative peaks is a technique which enables the required high degree of precision to be attained by averaging over many individual pulse waveforms. One possible realization is the "matched filter" concept described by Shapiro.[5] Another possibility is to use a sampling detector driven by sampling pulses with variable but known delay and recording the detected waveform as an average of many samples.

Even if the cycle detection system is not entirely safe on an absolute basis, it can be sufficiently useful if it allows retrieving the same cycle again and again on a given system and receiving site. The same point of lock can then be reproduced after interruptions of either the receiver or the local clock. If this is possible and if the whole system has been calibrated so that the total delay from the transmitter to the receiver output is known, the LORAN-C system can be used to refer the local clock to the UTC time scale with an uncertainty of about ± 0.1 μs. This is possible because of the established relations between the transmissions and the US Naval Observatory Time Scale. As a service to qualified users, the USNO publishes[8] tables indicating the times of coincidence (TOC) of the transmitted pulses with integer UTC seconds. The periods separating these TOCs depend obviously on the group repetition rate of the used chain. For the rates S0 and SS0, the TOC period is one second, whereas for the rate SS1 (GRI = 99 900 μs), one must wait for 999 seconds between two coincidences.

For daily routine operations, a 1 p.p.s. output derived from the 100 kHz signal locked (see Fig. 8.9) to the LORAN-C signal is very useful. A comparison with the local clock pulse by means of a time interval counter allows a fast daily system check and the reading (or printing) can be used directly for the record. This is easier than reading the strip chart and safer than relying on the phase shift pulse accumulating counters found in some commercial receivers.

The short review given here on the LORAN-C system's use for time and frequency comparison is far from exhaustive but the author hopes to have awakened some interest in this method, which, as it has reached a high level of sophistication, is accessible to a large number of users. The reader interested in further details will find them in the references and in the literature cited therein.

8.5. VLF-SYSTEMS

The very low frequency band (Band 4: 3–30 kHz) is characterized by its history of early long distance communication with stable and reliable signal propagation. The utility of phase measurements on VLF signals was recognized only after the second world war, starting with the pioneering work of J. A. Pierce,[9, 10, 11] suggesting a worldwide VLF navigation system under the name of "Radux", a forerunner of the actual OMEGA navigation system. Today, the part of the band between 10 and 14 kHz is allocated to radio-navigation systems among which OMEGA is most generally known. Frequencies below 10 kHz are not allocated to radio services. The channel from 19·95 to 20·05 kHz is allocated to standard frequency transmissions. The remainder of the band is allocated to fixed and maritime mobile radio-telegraphy communications. A comprehensive and recent review of the evolution and current state of VLF techniques has been written by E. R. Swanson and C. P. Kugel[9] of the US Naval Electronics Laboratory Center, San Diego, an institution known for its participation in the Omega development. The techniques of frequency and time comparison by means of VLF signals are characterized by the following properties:

Worldwide coverage is possible from a single transmitter of sufficient effective radiated power.

Over medium and long distances, the phase of the received wave shows a marked shift between day and night on the propagation path, typically of the order of 60 μs with a diurnal shape as shown in Fig. 8.4 for a north-south path (Hawaii to Alaska, Ref. 9).

During sunlight on the propagation path, the day to day variation of the phase is only a few microseconds on intercontinental distances.

Transmitter and antenna circuits must be relatively narrow-band. The radiation efficiency of antennas is low owing to their small size in comparison with the wavelength.

The basic method of using VLF transmissions for TFD is phase-time comparison of the received signal with the local standard frequency signal, using a phase comparison receiver as shown in Fig. 8.3. The somewhat higher uncertainty in the day to day phase-time values as compared to medium distance LF-CW systems leads to a corresponding larger probable error in the frequency comparison for a given period of observation.

Swanson[9] has indicated the following average values to be expected using signals of the OMEGA system:

Apparent frequency stability	Observation period in days
10^{-11}	6
10^{-12}	26
10^{-13}	133

Keeping in mind that VLF signals are the only means available in large parts of the world for standard frequency comparisons with not too expensive equipment, these values look very good. As long as there is no regular (i.e. not experimental) satellite service available on a continuous basis, all users of precise time and frequency located in the southern hemisphere must rely on VLF transmissions.

The technical data of currently operational VLF transmitters are given in the Appendix 8.2. Some additional information is given here on the Omega system:

The Omega system (for further details see Ref. 9) is a projected global radio-navigation system which has been partially deployed for about 8 years. The system, to be used mainly for hyperbolic navigation, is based on a network of 8 transmitters which transmit sequentially on three basic frequencies, namely, 10·6, 11·33 and 13·6 kHz.

TABLE 8.4. Basic OMEGA signal format

Segment	A	B	C	D	E	F	G	H
Duration (s)	0·9	1·0	1·1	1·2	1·1	0·9	1·2	1·0

Pauses: 0·2 s

Station	Transmitted basic frequencies (kHz)							
A	10·2	13·6	11.33					
B		10·2	13·6	11·33				
C			10·2	13·6	11·33			
D				10·2	13·6	11·33		
E					10·2	13·6	11·33	
F						10·2	13·6	11·33
G	11·33						10·2	13·6
H	13·6	11·33						10·2

The transmission sequence is much slower than that of the LORAN-C system since at these low frequencies the rising time of the signals is long (several tens of milliseconds, see Ref. 9).

The basic sequence repetition interval is 10 seconds and the basic format is given in Table 8.4 overleaf. As shown in more detail in Appendix 8.2, the following four stations are currently in operation:

Station	Location
A	Aldra, Norway
B	Trinidad, West Indies
C	Haiku, Hawaii, USA
D	Lamoure, North Dakota, USA

Other stations are under construction or being planned in the following locations:

Liberia (under construction, will replace Trinidad in 1976),
Argentina,
Australia,
La Reunion Island, and
Japan.

These stations will complete the global 8-station network.

Additional frequencies provided with a new special multifrequency timing system[9] are omitted from Table 8.4. OMEGA phase tracking receivers operate according to the same basic principle as shown in Fig. 8.3 but require a switching system synchronized to the signal format.

A great deal of research has been done over 10 years to provide not only phase information but also complete timing information through VLF systems.[9] The limits of envelope timing using the leading edge of the signal have been investigated by Guisset, Detrie and De Prins[12] and Swanson and Kugel.[9] These authors find an uncertainty of 100 to 200 μs depending on distance, propagation conditions and noise level. It has not been possible to identify repeatedly a particular zero crossing of the signal with satisfactory confidence.

A combination of LF envelope timing and VLF phase-time measurement to achieve cycle identification on the VLF and LF signals was suggested in 1969 by Bonanomi, Chaslain and Rentsch.[13] However, no results with this approach have been published since and it is uncertain whether efforts in this direction have been pursued.

Several attempts at cycle identification on VLF signals have been made using double or multiple frequency transmissions. The basic idea is to determine phase-time coincidences which occur periodically between (at least) two signals having (slightly) different frequencies. The time interval separating the coincidences is in principle equal to the beat period. On ideal signals, this looks indeed simple and obvious: a 50 Hz difference results in a period of 20 ms, i.e. even ordinary HF time signals would allow this ambiguity to be resolved. Unfortunately, the propagation medium creates difficulties and, as shown in the review of Ref. 9, unambiguous and repeatable VLF cycle identification is a difficult problem which, despite great efforts, cannot be considered as having been solved. Propagation effects are the main causes of difficulties in cycle identification. One of the major problems in the reception of signals from a distant transmitter is related to the good propagation properties (i.e. low attenuation) of VLF, namely the interference between signals coming in on different paths, one shorter direct, one longer around the globe. In such conditions the loss of whole cycles can occur.

8.6. TELEVISION TIMING METHODS

The various methods of time and frequency measurement and dissemination via radio waves reviewed in the preceding sections all use a relatively narrow bandwidth in the electromagnetic spectrum. The very low information content of a time signal does not require the occupation of wide channels exclusively for such a service for the sake of higher precision, indeed strong pressures demand a more economical use of the spectrum.

Fortunately, as we have already seen with the example of the LORAN-C system, there are systems and services in operation which allow the addition of timing information in a "piggy-back" fashion so as not to perturb or interfere with the intended main purpose of the service.

Television broadcast services which are rapidly extending to cover the whole world offer various possibilities for time comparison and TFD. Most of it is described in some detail in Ref. 1, especially the considerable effort made at the NBS Time and Frequency Division in Boulder, Colorado.

Two basic ways of using TV can be distinguished, namely *passive* time comparison methods and several kinds of *active* TFD.

The passive TV comparison method makes use of the synchronization pulses in the TV signal without interfering with the normal TV broadcast service. This method was introduced originally by Tolman, Ptáček, Souček and Stecher[14] in 1967 and was quickly adopted by others[1, 15, 16] and is now used in many countries as a simple and very precise means of clock comparison. The principle is illustrated in Figs 8.10. The most simple configuration in Fig. 8.10(a) consists of one single TV transmitter used to com-

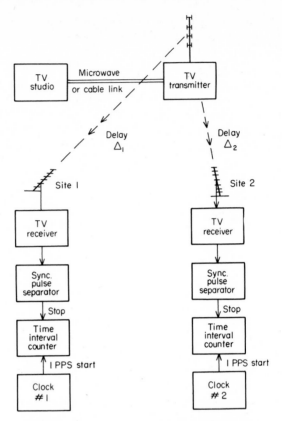

FIG. 8.10(a). Local level passive TV time comparison.

pare the times of two clocks (#1) and #2) located at sites where direct reception of the transmitter is possible. In the TV-studio, the synchronization
pulse pattern of the video signal is generated by using a stable crystal oscillator as the time base. In the receivers, circuits are added to separate one
particular synchronization pulse from the video signal after demodulation.
An easily identifiable pulse is chosen in the vertical blanking interval of the
TV signal. These pulses are not influenced by the varying content of the
image transmitting modulation. Which particular pulse selection is optimum
depends on the TV signal standard which in the USA is different from that
used in Europe. In the US standard the trailing edge of the sync pulse of line
No. 10 in the odd field is used.[1, 16] With the image repetition frequency of
30 per second used in the US, these pulses occur every $1/30$ s $= 33\cdot33$ milliseconds. In the CCIR 625-line standard used in Europe, the trailing edge of

the pulse of line No. 16 also in the odd field is selected.[15] The image repetition frequency is 25 per second and therefore, the pulse repetition interval is 40 ms.

These pulse repetition intervals of $33\frac{1}{3}$ or 40 ms determine the possible ambiguity of the clock time comparison, i.e. the clocks #1 and #2 must be synchronized by other means in order to make sure that the measurement at both clock sites is made on the same sync pulse transmitted. It is however easy to make this synchronization to one millisecond by means of either HF or LF time signals.

The measurement proceeds then in the following way: The 1 second pulse from a local clock starts a time interval counter. The next arriving TV sync pulse stops the counter, displays and prints the result. This measurement is repeated every second at both clock sites.

· The clock time differences are computed as follows: Let t be a common time, e.g. UTC. The transmitter sends its pulse at $t = T_s$. It gets to the respective clock sites and stops the counters at:

$$t = T_s + \Delta_1 \quad \text{(clock 1)}$$
$$t = T_s + \Delta_2 \quad \text{(clock 2)}$$
$$(8.3)$$

Δ_1 and Δ_2 are the propagation and equipment delays separating the instant when the sync pulse is sent from that when it stops the time interval counters. These were started by the preceding 1 p.p.s. pulses of clocks 1 and 2 at $t = T_1$ and $t = T_2$ respectively, whereby we assume that the clocks are nearly synchronous, i.e. $T_2 - T_1$ is not more than a few milliseconds, as mentioned before. The time interval counters therefore show the displays:

$$T_{d_1} = T_s + \Delta_1 - T_1 \text{ counter at clock #1}$$
$$T_{d_2} = T_s + \Delta_2 - T_2 \text{ counter at clock #2}$$
$$(8.4)$$

which are recorded (printed or punched, etc.).
The difference is:

$$T_{d_2} - T_{d_1} = \Delta_1 - \Delta_2 + T_2 - T_1 \quad (8.5)$$

and thus the clock time differences:

$$T_2 - T_1 = T_{d_1} - T_{d_2} + \Delta_2 - \Delta_1. \quad (8.6)$$

It is important to note that *provided the same TV pulse is stopping the counters* at both clock sites, the time of transmission (T_s) does not appear in the result any more and that the only calibration constant is the path delay difference $\Delta_2 - \Delta_1$. For line of sight reception, the path delay difference can easily be computed from the radial distance difference of the receiving antennas to the transmitter antenna with a precision better than 1 μs. The equipment delay differences can also be kept very small (tens of nanoseconds) if similar apparatus is used at the respective clock sites. Variations of the path delay difference are in most cases due to the equipment (especially if cheap commercial TV receivers are used) rather than to the radio path. A standard deviation of ± 50 ns is a typical order of magnitude which can be obtained even with simple equipment.

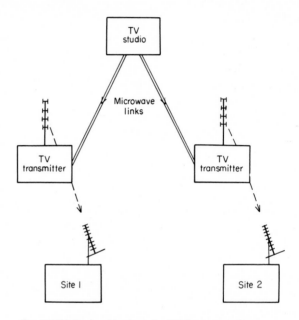

FIG. 8.10(b). Regional level passive TV time comparison.

Line of sight reception is limited to relatively small regions but the examples of the French atomic clock comparison system among various agencies in the suburbs of Paris are a good demonstration of the usefulness of this method.[15]

The passive TV comparison method can also be used, however, over much larger distances, namely wherever an extended and well-organized TV broadcasting network exists. The additional elements entering into the measurement system are the microwave links used for programme distribution. An example is given in Fig. 8.10(b).

The technique of measurement is the same as that outlined before, the only difference appearing in the values of Δ_1 and Δ_2 and, of course, the essential condition being that a common source (TV studio) is used for the measurement. This extended measurement technique has been used by the originators of the TV-method[14] and soon after on a very large scale between the US National Bureau of Standards and the US Naval Observatory.[16] The excellent results obtained over the long-distance link from Washington DC to Boulder (Colorado) are reproduced from Ref. 1 in Fig. 8.11. It is a plot of the Allan-variance σ_y versus measurement period τ computed from daily time comparisons over three different TV networks (NBC, National Broadcasting Co.; ABC, American Broadcasting Co. and CBS, Columbia Broadcasting System).

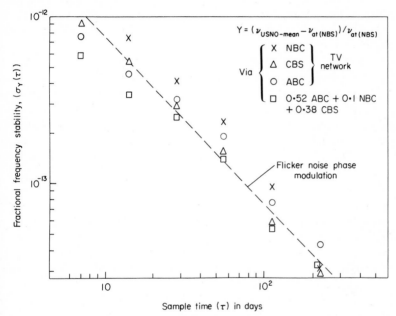

FIG. 8.11. Results obtained over the long-distance link from Washington DC to Boulder (Colorado).

The main problem in making TV-time comparisons over long distances covered by an extended broadcasting network is the close cooperation required between the network operating agency and the experimenters, in order to determine and control the path delay through the branches of the network used for the measurement. On a long distance path, the signals are routed over a large number of relay stations. As long as the configuration of the equipment is not changed, the path delay remains very constant (variations of less than 50 ns) but any changeover, e.g. to standby transmission equipment in case of failures changes the delay by a much larger amount.

Large changes in delay are also caused by rerouting of signals for various reasons, e.g. change of programme generating studio, etc.

Whereas local TV-time comparisons using a common transmitter do not require any cooperation on the part of the broadcast operating agency—it may even not be aware of the use made of its signals—long distance comparisons require some limited cooperation, namely information about the status of the network at the times when measurements are effected. In large networks, this information is often difficult to obtain even for the network operator. Under pressure from the hectic life of broadcast operations, he is usually satisfied if he gets his programme disseminated to all stations in good quality and has some difficulties in worrying about nanosecond delays which changed because in some relay station a switchover to standby equipment occurred.

Cooperation of a quite different nature, i.e. more active, is required from the broadcasting agencies for active TFD over the TV networks. Efforts in this direction have been initiated by the NBS since around 1970 and various possibilities of active TV-TFD are described in Ref. 1. Active TFD is addressed to a much larger class of users beyond the scientific community but obviously limited to the parts where it is implemented. Two possibilities of active TFD via television have been investigated and developed in some detail, namely:

(a) Standard frequency dissemination using either a 1 MHz burst in the vertical blanking interval or a stabilized colour-burst signal
(b) Insertion of a time code in the vertical blanking interval.

In the NTSC colour television standard used in the USA, a short burst of $5 \text{ MHz} \times \frac{63}{88} = 3 \cdot 57 \ldots \text{ MHz}$ is transmitted on each horizontal sync pulse. Stabilization of this frequency to a high degree was considered useful in improving the quality of the colour transmissions and was introduced a few years ago by the three TV networks. A PLL synthesizer has been developed by the NBS as an accessory to a colour TV receiver,[17] enabling frequency comparisons to be made to about 1 part in 10^{11} with 100 s measurement time.

1 MHz bursts have been included in the "NBS TV Time System" developed and demonstrated in 1971. This system is also the first attempt to disseminate precise time by means of a time code inserted in the line No. 1 (in the vertical blanking interval). The code carries hours, minutes and seconds information which can be displayed on the receiver's screen by means of a decoder and character generator. Despite successful demonstrations of the system, its general implementation has been delayed. In Europe, no plans are known for an eventual introduction of active time code systems in TV signals. It is

doubtful whether free lines in the vertical blanking interval could be made available for such a system exclusively since in the meantime, these free spaces have been discovered as a much more general means of dissemination of alphanumeric information. CEEFAX, ORACLE and TELETEXT (a combination of the two former systems) developed in Great Britain are examples of these new information broadcast facilities which may eventually include time.

One further example is the use of stabilized TV carrier frequencies as standard references in the VHF and UHF bands. In the Federal Republic of Germany efforts are being made to stabilize most TV transmitter carrier frequencies to 1 part in 10^{10} or better for the purpose of reducing mutual interference by means of the so called "precision offset" method. The use of TV signals for time and frequency measurements is a relatively recent field which remains open for further developments.

8.7. SATELLITE SYSTEMS

The use of satellite communications for frequency and time comparisons between distant clocks is almost as old as the artificial earth satellite technology itself. The first active satellite designed for telecommunications, TELSTAR, was used in 1962 for a transatlantic clock time comparison between the Royal Greenwich Observatory and the US Naval Observatory.[18] In this first satellite synchronization experiment, the excellent precision of about ± 1 μs was attained, using the two-way transmission method (see below).

In 1965 similar experiments were performed over the NASA satellite RELAY II between the US and Japan using refined techniques which improved the measurement precision to ± 0.1 μs.[19] The potential for high precision time comparisons offered by the wideband characteristics of transponders (transmitter–receivers) on board of satellites was thus demonstrated at an early time. Several other experiments have been performed since, some in connection with satellite navigation systems and others for evaluation of possible TFD systems.[1] No regular and permanent satellite TFD service has yet been installed. The general decrease in funds available for space activities and much closer scrutiny of budgets for matching the lesser funds to recognized needs explains the delay in introducing satellite TFD services.

The World Administrative Radio Conference for Space Telecommunications has allocated the following frequency bands for satellite TFD:

> 400·1 MHz $\pm$ 25 kHz
> 4202   MHz $\pm$ 2 MHz (space to Earth)
> 6427   MHz $\pm$ 2 MHz (Earth to space).

In view of the large bandwidths provided, these channels should offer interesting possibilities. Even the narrower UHF channel is 2·5 times wider than that available to LORAN-C (Section 8.4) and the SHF channels should allow very high precision work. As the allocations are not exclusive for the TFD service but are to be coordinated with radio-navigation satellite systems, much technical and administrative coordination work on a international level remains to be done, since satellite systems are worldwide by nature. The multitude of possible signal formats will make agreement on a standard signal a difficult task which might take many years to complete.

On the other hand, possibilities for time comparison using existing or future satellite navigation systems continue to be investigated and the interested reader should keep aware of the current publications in periodicals (see chapter 9) and conferences.

For time and frequency comparisons using satellites one of the main problems is to determine the propagation *path delay*. As mentioned above, the highest precision of measurement is obtained by using the so-called

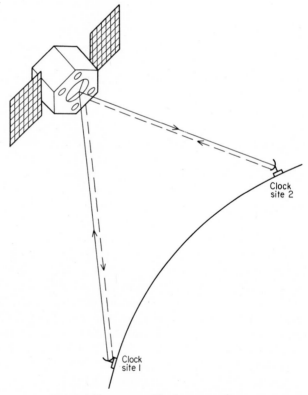

Fig. 8.12. Two-way satellite time comparison.

"*two-way*" *method*, illustrated in Fig. 8.12. This method requires a full duplex communication system including transmitters and receivers at each terminal. If the transmissions are made at the same time, it is possible to calibrate the path delay with high accuracy even on moving satellites. A possible way of doing the measurement is illustrated in the timing diagram of Fig. 8.13. We

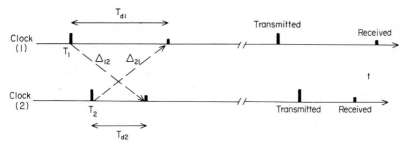

FIG. 8.13. Two way satellite time comparison—timing diagram

assume a common reference time scale t, e.g. UTC as in the example discussed in the preceding section. At each clock site, the local clock transmits a pulse at regular intervals, e.g. 1 p.p.s. At the instant of transmission the pulse starts a time interval counter. The pulse received from the partner station stops the counter. We have thus the following relations:

Clock (1) transmits at $t = T_1$
          receives at $t = T_2 + \Delta_{21}$

and the counter at Site 1 displays a time difference:

$$T_{d_1} = T_2 + \Delta_{21} - T_1. \tag{8.7a}$$

Clock (2) transmits at $t = T_2$
          receives at $t = T_1 + \Delta_{12}$

and the counter at Site 2 displays a time difference:

$$T_{d_2} = T_1 + \Delta_{12} - T_2 \tag{8.7b}$$

as illustrated in Fig. 8.13.

Δ_{12} and Δ_{21} are the total (sum of propagation and equipment) path delays from Site 1 to Site 2 and vice versa. Reciprocity, i.e. $\Delta_{12} = \Delta_{21}$ can be assumed if

(a) the satellite is truly geostationary; and
(b) the transmitter and receiver channel delays are identical or known and corrected for both stations.

From the relations 8.7a and b we obtain the two following results:
  The double path delay

$$\Delta_{12} + \Delta_{21} = T_{d_1} + T_{d_2} \tag{8.8}$$

and the time difference, referred to the reference time scale t:

$$T_2 - T_1 = \tfrac{1}{2}(T_{d_1} - T_{d_2}) + \tfrac{1}{2}(\Delta_{12} - \Delta_{21}). \tag{8.9}$$

This time difference result is not in compliance with the standard notation of Appendix 4.1 but easily transformed into clock reading differences as adopted by changing the sign; i.e. the difference between the clock readings R_1, R_2 at $t = T_1$ is

$$R_1 - R_2 = \tfrac{1}{2}(T_{d_2} - T_{d_1}) + \tfrac{1}{2}(\Delta_{21} - \Delta_{12}) + \Delta T_2 \tag{8.10}$$

where ΔT_2 is the small clock error accumulated by clock 2 in the short interval $T_1 - T_2$. This error is negligible in most cases.

Computation of the results at both stations requires that the counter readings must be exchanged. This can be done either by using the channel to transmit the result in digitally encoded form or by retransmitting another pulse with a known delay after the received pulse, so that both intervals T_{d_1}, T_{d_2} can be measured at one end of the link. The latter procedure has been used in the early experiments described in the Refs 18 and 19.

The precision of measurement is limited by the uncertainty of the arrival time measurement (depending on pulse shape and signal to noise ratio*) and above all by the precision of path delay compensation or correction. In (8.10), we see that the term due to path delay vanishes if the path is exactly reciprocal. This can be assumed or at least the error limits estimated by recording the double path delay of (8.8). The best precision is obtained if the two clocks are already close to synchronism, i.e. $T_2 \sim T_1$ or $R_1 \sim R_2$.

Knowledge of the satellite orbit elements is also useful as a cross-check, especially for medium or low altitude satellites which move relatively to the observers.

Path delay information which by the nature of the systems requires knowledge about the orbit and the position of the satellite (i.e. the satellites ephemerides) becomes quite important for the users of *one-way TFD systems* using satellites, i.e. the vast majority of possible users.

The various experiments described in Ref. 1 show that the precision of path delay corrections was usually of the order of 10 to 50 µs, depending on the quality of orbit information available. This order of magnitude applies to satellites which do not contain a very stable on-board clock but serve as moving relay stations to disseminate time from a ground-based clock.

---

* The measurement statistics can be improved by measuring at higher rates than 1 p.p.s. and averaging.[18, 19]

Much better results have been reported by R. L. Easton[20, 22, 23] using signals from the TIMATION II navigation satellite which contains a stable crystal controlled clock oscillator: A time transfer precision of 1 to 4 μs depending on quality and age of orbital data has been obtained. Two factors play in favour of higher precision in this case:

TIMATION II is a low altitude satellite allowing good signal to noise ratio; furthermore, it is a navigation satellite[21] and part of a system concept which relies on timing and on precise orbit data for its main purpose, i.e. position determination for navigators. Satellite navigation systems using time measurement techniques are currently in a phase of important developments as shown by the project called NAVSTAR-Global Positioning System (GPS).[24] In its ultimate development, planned for about 1984, this system will consist of a network of 24 satellites orbiting in three different planes, each inclined approximately 63° to the Equator and having an orbital period of 12 hours (i.e. medium altitude, 10000 nautical mile orbits). This configuration allows at least six satellites to be simultaneously in view from any point on the globe and at any time. Each satellite will have its own very stable clock on board and the system will be monitored and controlled from a network of ground stations.

The control parameters include tracking and updating of orbital data as well as data for the timing of the system. The satellites transmit pseudo-random "noise" (PRN) coded signals in the 1600 MHz band (allocated for radio-navigation) in a system bandwidth of 20 MHz.

By listening to four satellites, the user can determine his position in three dimensions as well as the system time. The system concept using accurate and stable clocks on board satellites also requires corrections to be made for relativistic effects. The largest and therefore the best known effect is the gravitational redshift* which, for a clock orbiting at the 12-h period altitude, is of the order of $y = 2.5 \times 10^{-10}$ referred to a clock on the Earth's surface.[25] Eccentricity of the orbit will cause a measurable frequency modulation which must be taken into account. These corrections can be handled together with the orbital data required in any case for the operation of the system. Its proponents claim a positioning accuracy in three dimensions of 10 m or better with 0·9 probability. To achieve this, relative timing precision alone must be within 30 ns or less, a rather severe requirement indeed. Many technical details of this project are still under development and it is somewhat premature to go into them here. The reason that this project has been mentioned in spite of its still early state of development is its tremendous potential as a truly global and accurate navigation system. It combines many ideas developed before on terrestrial systems and is actually one of the most interesting applications of time and frequency technology proposed.

* See Chapter 1. As seen from the Earth, actually a "blueshift" is observed.

Satellites transmitting signals which are well defined in time can be used for passive time transfer by applying the same principle as with TV or LORAN-C transmitters (see preceding sections). The only additional problems arise in the determination of the propagation delays which are variable even for a geostationary satellite. As mentioned above, navigation systems such as TIMATION and the planned NAVSTAR-GPS must deliver the necessary information for their own operation. Another probable future possibility for precise passive time comparison are geostationary direct TV broadcast satellites, which can be used within their areas of visibility in a similar way as a TV transmitter on a local basis provided that continuous and good data on the satellite's position are available.

From the above discussions the reader will notice some similarity between the terrestrial means of comparison and those using satellites concerning the two following points:

(a) *Navigation* systems relying on precise timing constitute an excellent means for *time comparison* even for users who are not directly interested in the navigation aspect.

(b) *Path delay* determination constitutes the main problem in precise time comparison between clocks located in distant sites.

We should not close this chapter without making another short remark on the problem of relativity. In Earth-based timing systems, relativistic corrections (e.g. gravitational redshift) are just beyond the accuracy capabilities of actual state of the art clocks and can therefore still be neglected except for a few high accuracy applications. In future systems using space vehicles and accurate on-board clocks, the problem of distinguishing between standard or proper time as generated by the clocks and coordinate time in the relativistic sense becomes significant. As these questions extend beyond the scope of this textbook devoted to measurements within the usual laboratory frames, we shall not elaborate further on them but refer the interested reader to a few recent texts on gravitation and timing.[26, 27, 28]

8.8. MISCELLANEOUS RADIO (AND OTHER) TIME COMPARISON SYSTEMS

As mentioned in the introduction to this chapter (Section 8.1) only those methods, which in the author's view are currently the most important ones could be reviewed in some detail. Many other approaches have been tested, some with great success but have not been reported here because of the limited available space. These methods are characterized by some restrictions limiting their general usefulness: geographical restrictions due to short signal range, limited performance, special expensive equipment required, availability limited in time and limited disclosure of pertinent system data are reasons for which we are giving only a list with certain bibliographic references.

A. *Other navigation systems*[1]
LORAN-A: Medium-frequency navigation system, using pulses on carrier frequencies around 1800 kHz.
LORAN-D: Variety of LORAN-C on 100 kHz, as yet geographically limited and temporary.

B. *Aircraft collision avoidance systems*[29]
Still under discussion and introduction in the proposed form very uncertain.

C. *Radio transmission, general*
VHF dissemination: Local importance only but good performance possible (see Appendix 8.2, list of stations.)
  Microwave links.
  Radio broadcasts: good potential with LF (long-wave) transmitters (see Appendix 8.2).
  Sub-audio phase modulated time-code (French proposal to CCIR 1974).

D. *Use of passive reflectors*
Moon-bounce experiments with radio[30] or optical signals.[31]
  Transmissions via Meteor Trails.[1]

E. *Pulsars*
Use of radio or optical pulsar signals for passive time transfer.[1]

F. *Aircraft flyover*
"Portable clock" trip plus VHF two-way comparison using airborne atomic clock.[1, 32]

G. *Power line phase*
The a.c. power line system can be used in principle for phase comparisons in a similar way to that for VLF transmissions but the attainable precision is rather low.[1]

The examples cited show that much imagination has been devoted to the investigation of time and frequency comparison techniques in the past. It can never be excluded that some of the methods just barely mentioned in this list could become of widespread use. Any potential user of time and frequency measurement techniques has a wide choice of methods available in principle. Which is the best one to use depends greatly on his particular condition and it is therefore impossible to give a general opinion. However, the author hopes to have given enough basic information on the various possibilities for the reader to evaluate those available to him in his particular case.

The very broad technical domain outlined in this chapter is still in evolution. We have seen much cross-fertilization between time and frequency and navigation systems technology which, with its latest trends, leads the four-dimensional space-time coordinate system from a theoretical concept to a practical reality.

Another field of extending future applications for time and frequency technology is in digital communications networks which are time ordered systems by definition. Many of the ceoncepts developed in time and frequency measurements will find useful applications there and digital communications systems will offer new means of time comparison and measurement.

REFERENCES

1. B. E. Blair. Time and frequency dissemination: An overview of principles and techniques. *In* "Time and Frequency: Theory and Fundamentals" (B. E. Blair, Ed.). NBS Monograph 140, pp. 233–314. *US Govt Printing Office, Washington, DC, May 1974*. (See Ref. 13 to Chapter 4.)
2. A. H. Morgan, "Precise Time Synchronization of Widely Separated Clocks", NBS Tech. Note 22, 65 pages (NTIS PB 151 381), July 1959.
3. G. Becker and G. Kramer. Methoden des internationalen Zeit- und Frequenzvergleichs mit Längstwellen. *Frequenz*, **9**, 256–261 (1969).
4. D. H. Andrews and J. De Prins. Reception of low frequency time signals. *Frequency*, **6** (9), 13–21 (1968).
5. L. D. Shapiro. Time synchronization from LORAN-C, *IEEE Spectrum*, **5** (8) 46–55 (1968).
6. C. E. Potts and B. E. Wieder. Precise time and frequency dissemination via the LORAN-C system. *Proc. IEEE*, **50** (5), 530–539 (1972).
7. J. Ph. Mellana. Swiss Post Office Internal Report VD22008 A (unpublished).
8. US Naval Observatory Time Service Announcements Series No. 9.
9. E. R. Swanson and C. P. Kugel. VLF timing: Conventional and modern techniques including Omega. *Proc. IEEE*, **60** (5), 540–551 (1972).
10. J. A. Pierce, A. T. Waterman Jr., and E. H. Cooper. "Time Phase Interference Between Modulated Ground and Sky Waves". Cruft Laboratory, Harvard University, Report 15, June 1947.
11. J. A. Pierce. "Radux". Cruft Laboratory, Harvard University, Report 17, July 1947.
12. J. L. Guisset, R. Detrie and J. De Prins. Réception de Signaux horaires sur ondes myriamétriques. *Bull. Classe Sci.* (*Acad. Roy. Belg.*), série 5, **52**, 490–499 (1966).
13. J. Bonanomi, Cl. Chaslain and E. Rentsch. "Le service horaire HBG, expériences

et résultats après 3 années d'exploitation". C. R. Colloque International de Chronométrie, pp. A23. 1–7. Paris, 1969.

14. J. Tolman, V. Ptáček, A. Souček and R. Stecher. Microsecond clock comparison by means of TV synchronizing pulses. *IEEE Trans. Instrum. Meas.*, IM-**16** (3) 247–254 (1967).

15. P. Parcelier. Time synchronization by television. *IEEE Trans. Instrum. Meas.*, IM-**19** (4) 233–238 (1970).

16. D. D. Davis, B. E. Blair and J. F. Barnaba. Long-term continental US timing system via television networks. *IEEE Spectrum*, **8** (8), 41–56 (1971).

17. D. D. Davis. Frequency standard hides in every color set. *Electronics*, **44** (18), 96–98 (1971).

18. J. McA. Steele, W. Markowitz and C. A. Lidback. Telstar time synchronization. *IEEE Trans. Instrumen. Meas.*, IM-**13** (4), 164–170 (1964).

19. W. Markowitz, C. A. Lidback, H. Uyeda and K. Muramatsu. Clock synchronization via Relay II satellite. *IEEE Trans. Instrum. Meas.*, IM-**15** (4), 177–184 (1966).

20. R. L. Easton. Timing receiver for Timation satellite. *In* Proc. 3rd Ann. PTTI Planning Meeting (US Naval Research Laboratory, Washington DC, 16–18 November 1971). USNO, Washington DC, pp. 45–66, 1972 (NTIS, AD 758739).

21. R. L. Easton. The role of time/frequency in navy navigation satellites, *Proc. IEEE*, Vol. **60** (5), 557–563 (1972).

22. R. L. Easton. British American satellite time transfer experiment. *In* "Proc. 4th Ann. PTTI Planning Meeting" (NASA Goddard Space Flight Center, Greenbelt, Md, 14–16 November 1972). NASA Doc. No. X-814–73–72, pp. 14–28.

23. R. L. Easton, H. M. Smith and P. Morgan. Submicrosecond time transfer between the United States, United Kingdom and Australia via satellite. *In* "Proc. 5th Ann. PTTI Planning Meeting" (NASA Goddard Space Flight Center, Greenbelt, Md, 4–6 December, 1973). NASA Doc. No. X-814-74-225, pp. 163–184.

24. B. W. Parkinson. NAVSTAR: Global Positioning System, an evolutionary research and development program. *In* Proc. 6th Ann. PTTI Planning Meeting. (US Naval Research Laboratory, Washington, D.C, 3–5 December, 1974). NASA Goddard Space Flight Center, Greenbelt Md, NASA Doc. No. X-814-75-117, pp. 465–495.

25. R. F. C. Vessot and M. Levine. Measurement of the gravitational redshift using a clock in an orbiting satellite. *In* "Proc. Conf. on Experimental Tests of Gravitation Theories" (Pasadena, 11–13 November 1970), pp. 54–64. JPL Technical Memorandum 33–499, Jet Propulsion Laboratory, California Institute of Technology, Pasadena, California.

26. N. Ashby. "An Earth Based Coordinate Clock Network", NBS Tech. Note 659. US Govt Printing Office Catalog No. C13.46: 659, April 1975.

27. V. Reinhardt. Relativistic effects of the rotation of the Earth on remote clock synchronization. *In* "Proc. 6th Ann. PTTI Planning Meeting" (US Naval Research Laboratory, Washington, DC, 3–5 December 1974). NASA Goddard Space Flight Center, Greenbelt Md, NASA Doc. No. X-814-75-117, pp. 395–424.

28. J. B. Thomas. A relativistic analysis of clock synchronization. Same as Ref. 27 above, pp. 425–440. (Also to be published in the *Astronomical Journal*.)

29. R. E. Perkinson and F. D. Watson. Airborne collision avoidance and other applications of frequency and time. *Proc. IEEE*, **60** (5), 572–578 (1972).

30. W. H. Higa. Time synchronization with lunar radar. *Proc. I.E.E.E.*, **60** (5), 552–557 (1972).

31. P. L. Bender, D. G. Currie, R. H. Dicke, D. H. Eckhardt, J. E. Faller, W. M. Kaula, J. D. Mulholland, H. H. Plotkin, S. K. Poultney, E. C. Silverberg, D. T. Wilkinson,

J. G. Williams, C. O. Alley. The lunar laser ranging experiment. *Science*, **182**, 229–238 (1973).

32. J. Besson. Comparison of national time standards by simple overflight. *IEEE Trans. Instrum. Meas.*, IM-**19** (4), 227–232 (1970).

APPENDIX 8.1

Official CCIR Band Designations

Definition: Band Number N extends from 0.3×10^N to 3×10^N Hz

Band number	Frequency range (lower limit inclusive, upper limit exclusive)	Abbreviation	Corresponding metric subdivision
4	3 to 30 kHz	VLF	Myriametric waves
5	30 to 300 kHz	LF	Kilometric waves
6	300 to 3000 kHz	MF	Hectometric waves
7	3 to 30 MHz	HF	Decametric waves
8	30 to 300 MHz	VHF	Metric waves
9	300 to 3000 MHz	UHF	Decimetric waves
10	3 to 30 GHz	SHF	Centimetric waves
11	30 to 300 GHz	EHF	Millimetric waves
12	300 to 3000 GHz or 3 THz	—	Decimillimetric waves

Symbols and prefixes: Hz = hertz
k = kilo (10^3)
M = mega (10^6)
G = giga (10^9)
T = tera (10^{12})

Source: Chapter II, Article 5, 112 Spa 2, Radio Regulations (Ed. 1968, Rev. 1971), published by the International Telecommunication Union.

APPENDIX 8.2

STANDARD-FREQUENCY AND TIME-SIGNAL EMISSIONS

RECOMMENDATION 460-1

(Question 1/7)

The CCIR,   (1970–1974)

CONSIDERING

(a) that the Administrative Radio Conference, Geneva, 1959, allocated the frequencies 20 kHz $\pm$ 0·05 kHz, 2·5 MHz $\pm$ 5 kHz (2·5 MHz $\pm$ 2 kHz in Region 1), 5 MHz $\pm$ 5 kHz, 10 MHz $\pm$ 5 kHz, 15 MHz $\pm$ 10 kHz, 20 MHz $\pm$ 10 kHz and 25 MHz $\pm$ 10 kHz to the standard-frequency and time-signal service, requesting the CCIR to study the question of establishing and operating a world-wide standard-frequency and time-signal service;

(b) that additional standard frequencies and time signals are emitted in other frequency bands;

(c) the provisions of Article 44, Section IV, of the Radio Regulations;

(d) the continuing need for close cooperation between Study Group 7 and the Inter-Governmental Maritime Consultative Organization (IMCO), the International Civil Aviation Organization (ICAO), the General Conference of Weights and Measures (CGPM), the Bureau International de l'Heure (BIH) and the concerned Unions of the International Council of Scientific Unions (ICSU);

(e) the desirability of maintaining world-wide coordination of standard-frequency and time-signal emissions;

(f) the need to disseminate standard frequencies and time signals in conformity with the second as defined by the 13th General Conference of Weights and Measures (1967);

(g) the continuing need to make Universal Time (UT) immediately available to an accuracy of one-tenth of a second;

UNANIMOUSLY RECOMMENDS

1. that all standard-frequency and time-signal emissions conform as closely as possible to Coordinated Universal Time (UTC) (see Annex I); that the time signals should not deviate from UTC by more than one millisecond; that the standard frequencies should not deviate by more than 1 part in 10^{10}, and that the time signals emitted from each transmitting station should bear a known relation to the phase of the carrier;

2. that all standard-frequency and time-signal emissions should contain information on the difference between UT1 and UTC (see Annexes I and II);

3. that this document be transmitted by the Director, CCIR, to all Administrations Members of the ITU, to IMCO, ICAO, the CGPM, the BIH, the International Union of Geodesy and Geophysics (IUGG), the International Union of Radio Science (URSI) and the International Astronomical Union (IAU);

4. that the standard-frequency and time-signal emissions should conform to RECOMMENDS 1 and 2 above as from 1 January 1975.*

ANNEX 1

TIME SCALES

A. Universal Time (UT)

In applications in which an imprecision of a few hundredths of a second cannot be tolerated, it is necessary to specify the form of UT which should be used:

UT0   is the mean solar time of the prime meridian obtained from direct astronomical observation;

UT1   is UT0 corrected for the effects of small movements of the Earth relative to the axis of rotation (polar variation);

UT2   is UT1 corrected for the effects of a small seasonal fluctuation in the rate of rotation of the Earth;

UT1   is used in this document, since it corresponds directly with the angular position of the Earth around its axis of diurnal rotation. GMT may be regarded as the general equivalent of UT.

B. International Atomic Time (TAI)

The international reference scale of atomic time (TAI), based on the second (SI), as realized at sea level, is formed by the Bureau International de l'Heure (BIH) on the basis of clock data supplied by cooperating establishments. It is in the form of a continuous scale, e.g. in days, hours, minutes and seconds from the origin 1 January 1958 (adopted by the CGPM 1971).

C. Coordinated Universal Time (UTC)

UTC is the time-scale maintained by the BIH which forms the basis of a coordinated dissemination of standard frequencies and time signals. It corresponds exactly in rate with TAI but differs from it by an integral number of seconds.

---

* Report 517 is valid until 1 January 1975; after that date the Report should be considered as cancelled.

The UTC scale is adjusted by the insertion or deletion of seconds (positive or negative leap-seconds) to ensure approximate agreement with UT1.

D. DUT1

The value of the predicted difference UT1-UTC, as disseminated with the time signals is denoted DUT1; thus DUT1 $\approx$ UT1 $-$ UTC. DUT1 may be regarded as a correction to be added to UTC to obtain a better approximation to UT1.

The values of DUT1 are given by the BIH in integral multiples of 0·1 s.

The following operational rules apply:

1. Tolerances

1.1 The magnitude of DUT1 should not exceed 0·8 s.

1.2 The departure of UTC from UT1 should not exceed $\pm 0·9$ s*.

1.3 The deviation of (UTC plus DUT1) from UT1 should not exceed $\pm 0·1$ s.

2. Leap-seconds

2.1 A positive or negative leap-second should be the last second of a UTC month, but first preference should be given to the end of December and June, and second preference to the end of March and September.

2.2 A positive leap-second begins at 23 h 59 m 60 s and ends at 0 h 0 m 0 s of the first day of the following month. In the case of a negative leap-second, 23 h 59 m 58 s will be followed one second later by 0 h 0 m 0 s of the first day of the following month (see Annex III).

2.3 The BIH should decide upon and announce the introduction of a leap-second, such an announcement to be made at least eight weeks in advance.

3. Value of DUT1

3.1 The BIH is requested to decide upon the value of DUT1 and its date of introduction and to circulate this information one month in advance.†

---

* The difference between the maximum value of DUT1 and the maximum departure of UTC from UT1 represents the allowable deviation of (UTC + DUT1) from UT1 and is a safeguard for the BIH against unpredictable changes in the rate of rotation of the Earth.

† In exceptional cases of sudden change in the rate of rotation of the Earth, the BIH may issue a correction not later than two weeks in advance of the date of its introduction.

3.2 Administrations and organizations should use the BIH value of DUT1 for standard-frequency and time-signal emissions, and are requested to circulate the information as widely as possible in periodicals, bulletins, etc.

3.3 Where DUT1 is disseminated by code, the code should be in accordance with the following principles (except § 3.5 below):
—the magnitude of DUT1 is specified by the number of emphasized second markers and the sign of DUT1 is specified by the position of the emphasized second markers with respect to the minute marker. The absence of emphasized markers indicates DUT1 = 0;
—the coded information should be emitted after each identified minute. Full details of the code are given in Annex II.

3.4 Alternatively, DUT1 may be given by voice or in Morse code.

3.5 DUT1 information primarily designed for, and used with, automatic decoding equipment may follow a different code but should be emitted after each identified minute.

3.6 In addition, UT1 − UTC may be given to the same or higher precision by other means, for example in Morse code or voice, by messages associated with maritime bulletins, weather forecasts, etc.; announcements of forthcoming leap-seconds may also be made by these methods.

3.7 The BIH is requested to continue to publish, in arrears, definitive values of the differences UT1 − UTC, UT2 − UTC.

ANNEX II

CODE FOR THE TRANSMISSION OF DUT1

A positive value of DUT1 will be indicated by emphasizing a number (n) of consecutive second markers following the minute marker from second marker one to second marker (n) inclusive; (n) being an integer from 1 to 8 inclusive.

$$DUT1 = (n \times 0.1)\,s$$

A negative value of DUT1 will be indicated by emphasizing a number (m) of consecutive second markers following the minute marker from second marker nine to second marker ($8 + m$) inclusive, (m) being an integer from 1 to 8 inclusive.

$$DUT1 = -(m \times 0.1)\,s$$

A zero value of DUT1 will be indicated by the absence of emphasized second markers.

The appropriate second markers may be emphasized, for example, by lengthening, doubling, splitting or tone modulation of the normal second markers.

Examples:

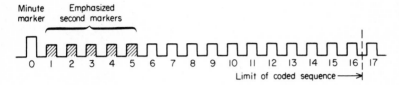

FIG. 1. DUT1 $= +0.5$ s.

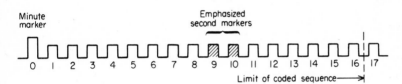

FIG. 2. DUT1 $= -0.2$ s.

ANNEX III

DATING OF EVENTS IN THE VICINITY OF A LEAP-SECOND

The dating of events in the vicinity of a leap-second shall be effected in the manner indicated in the following figures:

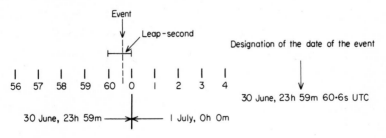

FIG. 3. Positive leap-second.

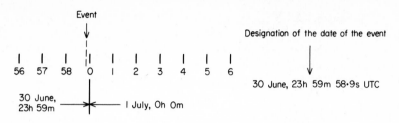

FIG. 4. Negative leap-second.

DRAFT REPORT 267–3(Rev. 76)

STANDARD FREQUENCIES AND TIME SIGNALS

Characteristics of standard-frequency and time-signal emissions in allocated bands and characteristics of stations emitting with regular schedules with stabilized frequencies, outside of allocated bands

(Question 1/7)

(1956–1959–1963–1966–1970–1974)

The characteristics of stations appearing in the following tables are valid as of 1 August 1974. For information concerning changes which may have occurred, reference may be made to the Annual Report of the Bureau International de l'Heure (BIH) or directly to the respective authority for each service as listed in Annex I.

TABLE I. Characteristics of standard-frequency and time

Station			Antenna(e)		Number of simultaneous transmissions	Period operat Days/Week
Call sign	Approximate location	Latitude Longitude	Type	Carrier power (kW)		
ATA	New Delhi, India	28°34′ N 77°19′ E	Horizontal dipole	2	2	6
BPV[14]	Shanghai, China	39°12′ N 121°26′ E	Omni-directional	5 to 15	2	7
FFH[1]	Paris, France	48°33′ N 02°34′ E	Vertical dipole	5	1	5[3]
IAM[1]	Roma, Italy	41°52′ N 12°27′ E	Vertical $\lambda/4$	1	1	6
IBF[1]	Torino, Italy	45°02′ N 07°46′ E	Vertical $\lambda/4$	5	1	7
JG2AR[1]	Tokyo, Japan	35°42′ N 139°31′ E	Omni-directional	3	1	1[3]
JJY[1]	Tokyo, Japan	35°42′ N 139°31′ E	Vertical $\lambda/2$ dipoles; ($\lambda/2$ dipole, top-loaded for 2·5 MHz)	2	4	7
LOL[1]	Buenos Aires, Argentina	34°37′ S 58°21′ W	Horizontal 3-wire folded dipole	2	3	7
MSF[1]	Rugby, United Kingdom	52°22′ N 01°11′ W	Horizontal quadrant dipoles; (vertical monopole, 2·5 MHz)	5	3	7

s in the allocated bands, valid as of 1 August 1974

d frequencies used	Duration of emission		Accuracy of frequency and time intervals (parts in 10^{10})	Method of DUT1 indication
Modulation (Hz)	Time signal (min)	Audio-modulation (min)		
1; 1000	continuous	4 in each 15	±5	
1[16] 1000[17]	19/30 (UTC)	10/30	±1	Direct emission of UT1 time signal[18]
1	continuous	nil	±0·2	CCIR code by lengthening to 0·1 s
1	continuous	nil	±0·1	CCIR code by double pulse
1	continuous	nil	±0·1	CCIR code by double pulse
nil	nil	nil	±0·5	No DUTI code
1[5] 1000[7]	continuous	27 in each 60	±0·5	CCIR code by lengthening
1; 440; 1000	continuous	3 in each 5	±0·2	CCIR code by lengthening
1	5 in each 10	nil	±0·02	CCIR code by double pulse

I

TABLE 1—*continued*

| Station | | | Antenna(e) | | Number of simultaneous transmissions | Peri ope |
| | | | | | | Days/week |
Call sign	Approximate location	Latitude Longitude	Type	Carrier power (kW)		
OMA[1]	Praha, Czechoslovak SR	50°07′ N 14°35′ E	T	1	1	7
RAT[1]	Moskva, USSR	55°19′ N 38°41′ E	Horizontal dipole	5	1	7
RCH[1]	Tashkent, USSR	41°19′ N 69°15′	Horizontal dipole	1	1	7
RID[1]	Irkutsk, USSR	52°46′ N 103°39′ E	Horizontal dipole	1	1	7
RIM[1]	Tashkent, USSR	41°19′ N 69°15′ E	Horizontal dipole	1	1	7
RKM[1]	Irkutsk, USSR	52°46′ N 103°39′ E	Horizontal dipole	1	1	7
RTA[1]	Novosibirsk, USSR	55°04′ N 82°58′ E	Horizontal dipole	5	1	7
RWM[1]	Moskva, USSR	55°19′ N 38°41′ E	Horizontal dipole	8	1	7
WWV[1]	Fort Collins, Colorado, USA	40°41′ N 105°02′ W	Vertical $\lambda/2$ dipoles	2·5 to 10	6	7

...rier ...Iz)	Modula-tion (Hz)	Time signal (min)	Audio-modulation (min)	Accuracy of frequency and time intervals (parts in 10^{10})	Method of DUT1 indication
5	1; 1000[8]	15 in each 30	4 in each 15	±10	
; 5	1; 10	39 in each 60	nil	±0·5	DUT1+dUT1:by Morse code each hour between minutes 11 and 12[12]
5	1; 10	38 in each 60	nil	±2	DUT1+dUT1:by Morse code each hour between minutes 51 and 52[12]
04; 004	1; 10	35 in each 60	nil	±0·5	DUT1+dUT1:by Morse code each hour between minutes 31 and 32[12]
10	1; 10	38 in each 60	nil	·	DUT1+dUT1:by Morse code each hour between minutes 51 and 52[12]
04; 004	1; 10	35 in each 60	nil	±0·5	DUT1+dUT1:by Morse code each hour between minutes 31 and 32[12]
96; 96; 996	1; 10	41 in each 60	nil	±0·5	DUT1+dUT1:by Morse code each hour between minutes 45 and 46[12]
15	1; 10	39 in each 60	nil	±0·5	DUT1+dUT1:by Morse code each hour between minutes 11 and 12[12]
; 5; 15; 25	1; 440; 500; 600	continuous [2]	continuous [10]	±0·1	CCIR code by double pulse. Additional information on UT1 corrections

TABLE 1—*continued*

Station			Antenna(e)		Number of simultaneous transmissions	Days/week	Perio opera
Call sign	Approximate location	Latitude Longitude	Type	Carrier power (kW)			
WWVH[1]	Kekaha, Kauai, Hawaii, USA	21°59′ N 159°46′ W	Vertical $\lambda/2$ dipole arrays	2·5 to 10	5	7	
WWVL[1] [9]	Fort Collins, Colorado, USA	40°41′ N 105°03′ W	Top-loaded vertical	1·8	[9]	[9]	
ZLFS	Lower Hutt, New Zealand	41°14′ S 174°55′ E		0·3	1	1	
ZUO[1]	Olifantsfontein, Republic of South Africa	24°58′ S 28°14′ E	Vertical monopole	4	1	7	2

Notes to Table I

The daily transmission schedule and hourly modulation schedule is given, where appropriate, in the form of Figs 1 and 2 supplemented by the following notes:

[1] These stations have indicated that they follow the UTC system as specified in Recommendation 460-1. Since 1 January 1972 the frequency offset has been eliminated and the time signals remain within about 0·8 s of UT1 by means of occasional 1 s steps as directed by the Bureau International de l'Heure.

[2] In addition to other timing signals and time announcements, a modified IRIG-H time code is produced at a 1-p.p.s. rate and radiated continuously on a 100 Hz sub-carrier on all frequencies. A complete code frame is 1 minute. The 100 Hz sub-carrier is synchronous with the code pulses, so that 10 ms resolution is obtained. The code contains DUT1 values and UTC time-of-year information in minutes, hours and days of the year.

[3] Each Monday.

[4] From 0530 to 0730 hours UT.

[5] Interrupted from 25 to 34 minutes of each hour.

[6] Pulse consists of 8 cycles of 1600 Hz tone. First pulse of each minute preceded by 655 ms of 600 Hz tone.

[7] 1000 Hz tone modulation between the minutes of 0–10, 20–25, 34–35, 40–50 and 59–60 except 40 ms before and after each second's pulse.

[8] In the period from 1800–0600 hours UT, audio-frequency modulation is replaced by time signals.

[9] Effective 1 July 1972, regularly scheduled transmissions from WWVL were discontinued.

dard frequencies used		Duration of emission		Accuracy of frequency and time intervals (parts in 10^{10})	Method of DUT1 indication
·ier ·lz)	Modula- tion (Hz)	Time signal (min)	Audio- modulation (min)		
5 15;	1; 440; 500; 600	continuous [2]	continuous [10]	±0·1	CCIR code by double pulse. Additional inform- ation on UT1 corrections
2	nil	nil	nil	±0·1	
5	nil	nil	nil	±1	
5	1	continuous	nil	±0·1	CCIR code by lengthening

Since that date, this station has been broadcasting experimental programmes on an inter-
mittent basis only.
[10] Except for voice announcement periods and the 5-minute semi-silent period each hour.
[11] 2·5 MHz; from 1800–0400 hours UT; 5 MHz; continuous.
[12] The information about the value and the sign of the DUT1 + dUT1 difference is transmitted
after each minute signal by the marking of the corresponding second signals by additional
impulses. In addition, it is transmitted in Morse code as indicated. UT1 information is
transmitted in accordance with CCIR code. Additional information dUT1 is given, specifying more precisely the difference UT1 − UTC down to multiples of 0·02 s, the total value of
the correction being DUT1 + dUT1. Positive values of dUT1 are transmitted by the marking of p second markers within the range between the 20th and 25th seconds so that dUT1
$= +0·02$ s $× p$. Negative values of dUT1 are transmitted by the marking of q second
markers within the range between the 35th and the 40th second, so that dUT1 $= -0·02$ s $× q$.
[13] 0330 to 1130 hours UT.
[14] Call sign in Morse.
[15] 15 MHz: from 0100 to 1600 hours UT, 5 MHz; from 1600 to 0100 hours UT,
10 MHz: continuous.
[16] Pulse consists of 10 cycles of 1000 Hz tone (10 ms width). First second of each minute consists of nine 10 ms pulses of 1000 Hz tone, the beginning of the first pulse indicate the minute.
[17] 1000 Hz tone modulation (950 ms) and standard interval (50 ms). First second of each minute
is similar with that of 10 ms time signal.
[18] 100 ms pulse of 1000 Hz tone, minute pulse lengthened to 500 ms. Emission between the
minutes of 10–15 and 40–45 each hour.

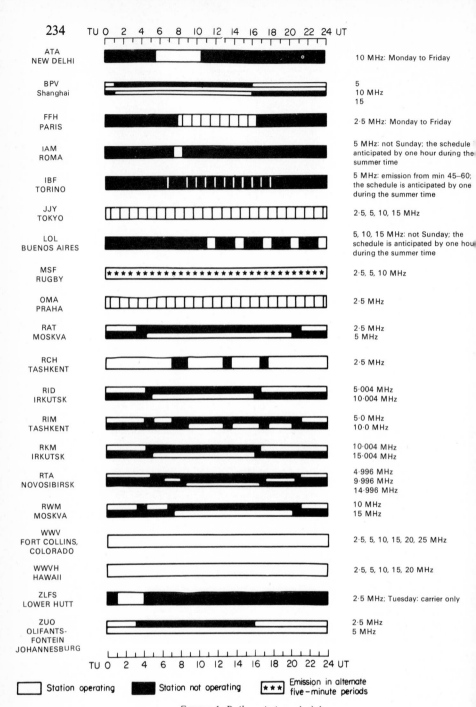

FIGURE 1. *Daily emission schedule.*

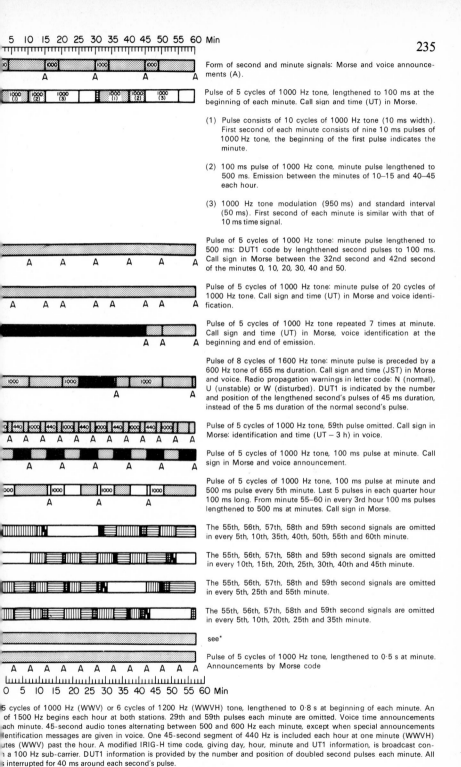

FIGURE 2.

TABLE II. Characteristics of standard-frequency and time-signal em

Station			Antenna(e)		Number of simultaneous transmissions	Peri oper
Call sign	Approximate location	Latitude Longitude	Type	Carrier power (kW)		Days/week
	Allouis, France	47°10′ N 02°12′ E	Omni-directional	1000 to 2000	1	7
CHU[1]	Ottawa, Canada	45°18′ N 75°45′ W	Omni-directional	3; 10; 3	3	7
	Donebach, FR of Germany	49°34′ N 09°11′ E	Omni-directional	250	1	7
DCF77[1]	Mainflingen, FR of Germany	50°01′ N 09°00′ E	Omni-directional	38 (2)	1	7
	Droitwich, United Kingdom	52° 16′ N 02° 09′ W	T	400	1	7
GBR[1][31]	Rugby, United Kingdom	52°22′ N 01°11′ W	Omni-directional	750 60[2]	1	7
HBG[39]	Prangins, Switzerland	46°24′ N 06°15′ E	Omni-directional	20	1	7
JJF-2[1] JG2AS	Kemigawa, Chiba Japan	35°38′ N 140°04′ E	Omni-directional	10	1	7 [5]
MSF	Rugby, United Kingdom	52°22′ N 01°11′ W	Omni-directional	25[2]	1	7

nal bands, valid as of 1 March 1976

ard frequencies used		Duration of emission		Accuracy of frequency and time intervals (parts in 10^{10})	Method of DUT1 indication
er z)	Modulation (Hz)	Time signal (min)	Audio-modulation (min)		
[33]	nil	nil	continuous A3	±0·5	
30; 35; 70	1[4]	continuous	nil	±0·05	CCIR code by split pulses
51	nil	nil	continuous A3	±0·05	
7·5	1	continuous [6]	continuous[7]	±0·005	CCIR code by lengthening to 0·2 s
00	nil	nil	A3 broadcast continuously	±0·2	
95 00	1[9]	4 × 5[10] per day	nil	±0·02	CCIR code by double pulse
5	1[28]	continuous[29]	nil	±0·2	No DUT1 transmission
)	1[21]	continuous [29]	nil	±0·5	
)	1[12]	continuous	nil	±0·02	CCIR code by double pulse

TABLE II—*continued*

Station			Antenna(e)		Number of simultaneous transmissions	Period operal	
Call sign	Approximate location	Latitude Longitude	Type	Carrier power (kW)		Days/week	
NAA[1][22] [32]	Cutler, Maine, USA	44°38′ N 67°17′ W	Omni-directional	2000 1000 [2]	1	7	
NBA[1][22] [32]	Balboa, Panama Canal Zone, USA	09°03′ N 79°39′ W	Omni-directional	300 110[2]	1	7	2
NDT[1][32]	Yosami, Japan	34°58′ N 137°01′ E	Omni-directional	40[2]	1	7	
NLK[1][32]	Jim Creek Washington, USA	48°12′ N 121°55′ W	Omni-directional	1200 130[2]	1	7	2
NPM[1][22] [32]	Lualualei, Hawaii, USA	21°25′ N 158°09′ W	Omni-directional	1000 630[2]	1	7	2
NSS[1][32]	Annapolis, Maryland, USA	38°59′ N 76°27′ W	Omni-directional	1000 400[2]	1	7	2
NWC[1][22] [32]	North West, Cape, Australia	21°49′ S 114°10′ E	Omni-directional	1000[2]	1	7	2
OMA	Podebrady, Czechoslovak SR	50° 08′ N 15° 08′ E	T	5	1	7	
RBU[1]	Moskva, USSR	55° 19′ N 38° 41′ E	Horizontal dipole	10	1	7	

rd frequencies used	Duration of emission		Accuracy of frequency and time intervals (parts in 10^{10})	Method of DUT1 indication
Modulation (Hz)	Time signal (min)	Audio-modulation (min)		
nil	nil	nil	± 0.1	
nil	[38]	nil	± 0.1	By Morse code each minute between s 56–59
nil	nil	nil	± 0.1	
nil	nil	nil	± 0.1	
nil	nil	nil	± 0.1	
nil	[27]	nil	± 0.1	By Morse code each minute between s 56–59
nil	[30]	nil	± 0.1	By Morse code each minute between s 56–58 [40]
1[9]	23 hours per day[16]	nil	± 10	No DUT1 transmission
1; 10	6 in each 60	nil	± 0.5	By Morse code each hour between minutes 6 and 7[41]

TABLE II—*continued*

Station			Antenna(e)		Number of simultaneous transmissions	Peri• oper•
Call sign	Approximate location	Latitude Longitude	Type	Carrier power (kW)		Days/Week
RTZ[1]	Irkutsk, USSR	52° 18′ N 104° 18′ E	Horizontal dipole	10	1	7
RV/166	Irkutsk, USSR	52° 18′ N 104° 18′ E	Horizontal dipole	40	1	7
SAZ[21]	Enkoping, Sweden	59° 35′ N 17° 08′ E	Yagi (12 dB)	0·1 (ERP)	1	7
SAJ[21]	Stockholm, Sweden	59° 20′ N 18° 03′ E	Omni-directional	0·06 (ERP)	1	1 [18]
VNG[1]	Lyndhurst, Victoria, Australia	38° 03′ S 145° 16′ E	Omni-directional	10	2	7
WWVB[1]	Fort Collins, Colorado, USA	40° 40′ N 105° 03′ W	Top-loaded vertical	13[2]	1	7
ZUO[26]	Olifantsfontein, Republic of South Africa	24° 58′ S 28° 14′ E	Omni-directional	0·08	1	7

Notes to Table II

[1] These stations have indicated that they follow one of the systems referred to in Recommendation 460-1.

[2] Figures give the estimated *radiated* power.

[3] Time code used which reduces carrier by 10 dB at the beginning of each second.

[4] Pulses of 300 cycles of 1000 Hz tone; the first pulse in each minute is prolonged.

[5] From Monday to Saturday, for JG-2AS.

ard frequencies used		Duration of emission		Accuracy of frequency and time intervals (parts in 10^{10})	Method of DUT1 indication
ier ?)	Modula-tion (Hz)	Time signal (min)	Audio-modulation (min)		
	1; 10	5 in each 60	nil	$\pm 0\cdot5$	By Morse code each hour between minutes 6 and 7[41]
		nil	broadcast	$\pm 0\cdot5$	
0	nil	nil	nil	± 50	
0	nil	nil	10[20]	± 1	
	1; 1000[24]	continuous	nil	± 1	CCIR code by 45 cycles of 900 Hz immediately following the normal second markers
	1[3]	continuous	nil	$\pm 0\cdot1$	No CCIR code
0	1	continuous	nil	$\pm 0\cdot1$	CCIR code by lengthening

[6] A1 time signals (interruptions of the carrier during 100 ms at the beginning of each second except for the second No. 59 of each minute) from the Physikalisch-Technische Bundesan-stalt. Since 6 June 1973, the number of the minute, hour, calendar day, calendar month and calendar year, as well as the day of the week, every minute beginning with the 20th second and ending with the 58th second, are transmitted in accordance with the official Time Scale CET (PTB) = UTC (PTB) + 1 h, in BCD code. A 0·1 s wide second marker is equivalent to "binary 0" and a 0·2 wide marker to "binary 1".

[7] Call sign is given by modulation of the carrier with 250 Hz tone three times every hour at the minutes 19, 39 and 59, without interruption of the time signal sequence.

[8] Maintenance period from 1300 to 1430 hours UT each day.

[9] A1 telegraphy signals.

[10] From 0255 to 0300, 0855 to 0900, 1455 to 1500 and 2055 to 2100 hours UT.

[11] Maintenance period from 1300 to 1600 UT on the first Sunday of each month.

[12] Carrier interrupted for 100 ms at each second and 500 ms at each minute; from 1430 to 1530 hours UT, A2 pulses are transmitted in the same form as for MSF 2·5, 5 and 10 MHz.

[13] Time pulses occur in groups of 8, one millisecond apart; 20 groups per second.

[14] Except from 1200 to 1800 hours UT each Tuesday. Each Wednesday the transmitter will operate at half power from 1200 to 2000 hours UT.

[15] Except from 1700 hours UT, Monday to 0200 hours UT, Tuesday on 1st and 3rd Mondays of each month.

[16] From 1000 to 1100 hours UT, transmission without keying except for call-sign OMA at the beginning of each quarter-hour.

[17] JJF-2: telegraph, JG2AS: from 2330 to 0800 hours UT.

[18] Each Friday.

[19] From 0930 to 1130 hours UT.

[20] 5 minutes at the beginning and 5 minutes at the end of the transmission for identification purposes only.

[21] Emission of the carrier of 500 ms duration at the beginning of each second where the 59th pulse is omitted each minute.

[22] FKS is used. Phase stable on assigned frequency.

[23] 4500 kHz, from 0945 hours UT to 2130 hours UT, 12000 kHz, from 2145 hours UT to 0930 hours UT, 7500 kHz, continuous service, with a technical interruption from 2230 hours UT to 2245 hours UT.

[24] Pulses of 50 cycles of 1000 Hz tone, shortened to 5 cycles from the 55th to the 58th second; the 59th pulse is omitted. At the 5th, 10th, 15th, etc. minutes, pulses from the 50th to the 58th second are shortened to 5 cycles; voice identification between the 20th and 50th pulses in the 15th, 30th, 45th and 60th minutes.

[25] Except first minute of each hour.

[26] Transmitter phase modulated; time signals and announcements as for ZUO 2·5 and 5 MHz (see Table I).

[27] Transmissions are temporarily suspended. FSK time signals are planned when transmissions are resumed.

[28] Interruption of the carrier during 100 ms at the beginning of each second; double pulse each minute; triple pulse each hour; quadruple pulse every 12 hours.

[29] In absence of telegraph traffic.

[30] Time signals on FSK during 2 minutes preceding 0030, 0430, 0830, 1230, 1630 and 2030 hours UTC.

[31] FSK is used, alternatively with CW; both carriers are frequency controlled.

[32] This station is primarily for communication purposes; while these data are subject to change, the changes are announced in advance to interested users by the U.S. Naval Observatory, Washington, DC, USA.

[33] Temporary.

[34] Except from 1400 to 1800 hours UT each Friday. Each Wednesday and Thursday the transmitter will operate at half power from 1200 to 2000 hours UT.

[35] Except from 1700 to 2200 hours UT 1st and 3rd Thursday of each month.

[36] Except from 1300 to 1900 hours UT each Wednesday.

[37] Except from 0000 to 0300 hours UT each Monday.

[38] Time signal on FSK 5 minutes before each even hour except 2355 to 2400 hours UT.

[39] Coordinated time signals.

[40] *DUT1 information in CCIR code*

dUT1 information. This additional information specifies more precisely the difference $UT1 - UTC$ down to multiples of 0.02 s, the total value of the correction being DUT1 $+ dUT1$.

A positive value of dUT1 is indicated by doubling a number (p) of consecutive second markers from second marker 21 to second marker $(20 + p)$ inclusive; (p) being an integer from 1 to 5 inclusive

$$dUT1 = p \times 0.02 \text{ s}$$

A negative value of dUT1 is indicated by doubling a number (q) of consecutive second markers following the minute marker from second marker 31 to second marker $(30 + q)$ inclusive; (q) being an integer from 1 to 5 inclusive

$$dUT1 = -(q \times 0.02) \text{ s}$$

The second marker 28 following the minute marker is doubled as parity bit, if the value of (p) or (q) is an even number, or if $dUT1 = 0$.

[41] The information about the value and the sign of the DUT1 + dUT1 difference is transmitted after each minute signal by the marking of the corresponding second signals by additional impulses. In addition, it is transmitted in Morse code as indicated. UT1 information is transmitted in accordance with CCIR code. Additional information dUT1 is given, specifying more precisely the difference $UT1 - UTC$ down to multiples of 0.02 s, the total value of the correction being DUT1 + dUT1. Positive values of dUT1 are transmitted by the marking of p second markers within the range between the 20th and 25th seconds so that dUT1 $= +0.02 \text{ s} \times p$. Negative values of DUT1 are transmitted by the marking of q second markers within the range between the 35th and the 40th second, so that $dUT1 = -0.02 \text{ s} \times q$.

TABLE III. Characteristics of

Station			Antenna(e)	
Call sign	Approximate location	Latitude longitude	Type	Ca po (k
LORAN-C SS7-M (9930-M)	Carolina Beach, N.C., USA	34° 03·8′ N 77° 54·8′ W	Omni-directional	70
LORAN-C SS7-W (9930-W)	Jupiter, Florida, USA	27° 02·0′ N 80° 06·9′ W	Omni-directional	30
LORAN-C[6] SS7-X (9930-X)	Cape Race, Newfoundland	46° 46·5′ N 53° 10·5′ W	Omni-directional	180
LORAN-C SS7-Y (9930-Y)	Nantucket Island, USA	41° 15·2′ N 69° 58·6′ W	Omni-directional	30
LORAN-C SS7-Z (9930-Z)	Dana, Indiana, USA	39° 51·1′ N 87° 29·2′ W	Omni-directional	40
LORAN-C SL7-M (7930-M)	Angissq, Greenland	59° 59·3′ N 45° 10·4′ W	Omni-directional	100
LORAN-C[6] SL3-M (7970-M)	Ejde, Faroe Is.	62° 18·0′ N 7° 04·5′ W	Omni-directional	40
LORAN-C SL3-W (7970-W)	Sylt, F.R. of Germany	54° 48·5′ N 8° 17·6′ E	Omni-directional	30
LORAN-C SL3-X (7970-X)	Boe, Norway	68° 38·1′ N 14° 27·1′ E	Omni-directional	20
LORAN-C[6] SL3-Y (7970-Y)	Sandur, Iceland	64° 54·4′ N 23° 55·3′ W	Omni-directional	180
LORAN-C SL3-Z (7970-Z)	Jan Mayen, Norway	70° 54·9′ N 8° 44·0′ W	Omni-directional	20
LORAN-C SL1-M (7990-M)	Simeri Crichi, Italy	38° 52·3′ N 16° 43·1′ E	Omni-directional	20

tional aids, valid as of 1 March 1976

Period of operation		Standard frequencies used		Duration of emission		Accuracy of frequency and time intervals (parts in 10^{10})
Days/Week	Hours/day	Carrier (kHz)	Pulse repetition in micro-seconds	Time signal	Audio-modulation	
7	24	100	[1] 99 300	continuous[3]	nil	±0·01
7	24	100	[1] 99 300	continuous[5]	nil	±0·01
7	24	100	[1] 99 300	continuous[5]	nil	±0·01
7	24	100	[1] 99 300	continuous[5]	nil	±0·01
7	24	100	[1] 99 300	continuous[5]	nil	±0·01
7	24	100	[1] 79 300	continuous[5]	nil	±0·01
7	24	100	[1] 79 700	continuous[5]	nil	±0·01
7	24	100	[1] 79 700	continuous[5]	nil	±0·01
7	24	100	[1] 79 700	continuous[5]	nil	±0·01
7	24	100	[1] 79 700	continuous[5]	nil	±0·01
7	24	100	[1] 79 700	continuous[5]	nil	±0·01
7	24	100	[1] 79 900	continuous[5]	nil	±0·01

TABLE III—*continued*

Station			Antenna(e)	
Call sign	Approximate location	Latitude Longitude	Type	Ca p((k
LORAN-C SL1-X (7990-X)	Lampedusa, Italy	35° 31·3′ N 12° 31·5′ E	Omni-directional	4
LORAN-C SL1-Y (7990-Y)	Targabarun, Turkey	40° 58·3′ N 27° 52·0′ E	Omni-directional	2
LORAN-C SL1-Z (7990-Z)	Estartit, Spain	42° 03·6′ N 3° 12·3′ E	Omni-directional	2
LORAN-C S1-M (4990-M)	Johnston Is.	16° 44·7′ N 169° 30·5′ W	Omni-directional	3
LORAN-C S1-X (4990-X)	Upolo Pt., Hawaii, USA	20° 14·8′ N 155° 53·1′ W	Omni-directional	3
LORAN-C S1-Y (4990-Y)	Kure, Hawaii, USA	23° 23·7′ N 178° 17·5′ W	Omni-directional	3
LORAN-C SS3-M (9970-M)	Iwo Jima, Japan	24° 48·1′ N 141° 19·5′ E	Omni-directional	40
LORAN-C SS3-W (9970-W)	Marcus Is., Japan	24° 17·1′ N 153° 58·9′ E	Omni-directional	40
LORAN-C SS3-X (9970-X)	Hokkaido, Japan	42° 44·6′ N 143° 43·2′ E	Omni-directional	4
LORAN-C SS3-Y (9970-Y)	Gesashi, Okinawa, Japan	26° 36·4′ N 128° 08·9′ E	Omni-directional	4
LORAN-C SS3-Z (9970-Z)	Yap, Caroline Is.	9° 32·8′ N 138° 09·9′ E	Omni-directional	400
LORAN-C SH7-M (5930-M)	St. Paul Pribiloff Is., Alaska	59° 09·2′ N 170° 15·0′ W	Omni-directional	3

Period of operation		Standard frequencies used		Duration of emission		Accuracy of frequency and time intervals (parts in 10^{10})
Days/week	Hours/day	Carrier (kHz)	Pulse repetition in micro-seconds	Time signal	Audio-modulation	
7	24	100	[1] 79 900	continuous[5]	nil	±0·01
7	4	100	[1] 79 900	continuous[5]	nil	±0·01
7	24	100	[1] 79 900	continuous[5]	nil	±0·01
7	24	100	[1] 49 900	continuous[5]	nil	±0·01
7	24	100	[1] 49 900	continuous[5]	nil	±0·01
7	24	100	[1] 49 900	continuous[5]	nil	±0·01
7	24	100	[1] 99 700	continuous[5]	nil	±0·01
7	24	100	[1] 99 700	continuous[5]	nil	±0·01
7	24	100	[1] 99 700	continuous[5]	nil	±0·01
7	24	100	[1] 99 700	continuous[5]	nil	±0·01
7	24	100	[1] 99 700	continuous[5]	nil	±0·01
7	24	100	[1] 59 300	continuous[5]	nil	±0·01

TABLE III—*continued*

Station			Antenna(e)	
Call sign	Approximate location	Latitude Longitude	Type	Carri pow (kW
LORAN-C SH7-X (5930-X)	Atta, Alaska	52° 49·7′ N 173° 10·9′ E	Omni-directional	300[
LORAN-C SH7-Y (5930-Y)	Pt. Clarence, Alaska	65° 14·7′ N 166° 53·2′ W	Omni-directional	1000[
LORAN-C SH7-Z (5930-Z)	Sitkinak, Alaska	56° 32·3′ N 154° 07·8′ W	Omni-directional	300[
OMEGA Ω/N	Aldra, Norway	66° 25′ N 13° 09′ E	Omni-directional	4[
OMEGA Ω/ND	Lamoure, North Dakota, USA	46° 22′ N 98° 21′ W	Omni-directional	10[
OMEGA Ω/T	Trinidad, West Indies	10° 42′ N 31° 38′ W	Omni-directional	1
OMEGA Ω/H	Haiku, Hawaii, USA	21° 24′ N 157° 50′ W	Omni-directional	2
OMEGA Ω/J	Tsuchima Is., Japan	34° 37′ N 129° 27′ E	Omni-directional	10

lses appear in groups of 9 for the master station (M) and groups of 8 for the secondary
(W X Y Z).
e the estimated radiated power.
r.
power.

Period of operation		Standard frequencies used		Duration of emission		Accuracy of frequency and time intervals (parts in 10^{10})
Days/week	Hours/day	Carrier (kHz)	Pulse repetition in micro-seconds	Time signal	Audio-modulation	
7	24	100	[1] 59 300	continuous[5]	nil	±0·01
7	24	100	[1] 59 300	continuous[5]	nil	±0·01
7	24	100	[1] 59 300	continuous[5]	nil	±0·01
7	24	10·2-A[3] 11$\frac{1}{3}$-C 13·6-B	nil	[3]	nil	±0·05
7	24	10·2-D[3] 11$\frac{1}{3}$-F 13·6-E	nil	[3]	nil	±0·01
7	24	10·2-B[3] 11$\frac{1}{3}$-D 13·6-C	nil	[3]	nil	±0·01
7	24	10·2-C[3] 11$\frac{1}{3}$-E 13·6-D	nil	[3]	nil	±0·01
7	24	10·2-H[3] 11$\frac{1}{3}$-B 13·6-A	nil	[3]	nil	±0·01

[5] Maintained within ±5 μs of UTC. Time of Coincidence (TOC) with the UTC second changes with the occurrence of leap-seconds and is designated in TOC Tables issued to interested users by the US Naval Observatory, Washington DC, USA.

[6] Dual-rated stations also transmitting on rate SL7 with a pulse repetition period of 79 300 microseconds.

TABLE IV. OMEGA signal format

	0·0s	1·0	2·0	3·0	4·0	5·0	6·0	7·0	8·0	9·0	10·0
Segment	A	B	C	D	E	F	G	H	
Duration	0·9	1·0	1·1	1·2	1·1	0·9	1·2	1·0	
KHz:									
10·2	Norway	Trinidad	Hawaii	North Dakota					
11⅓			Norway	Trinidad	Hawaii	North Dakota			
13·6		Norway	Trinidad	Hawaii	North Dakota				

Note 1.—Segment A does not begin at 0·0 second UTC. Time of segments changes with leap-seconds. Segment A begins at 58·0 seconds in January 1973.

Note 2.—The OMEGA stations are primarily for navigation purposes: while these data are subject to change, the changes are announced in advance to interested users by the US Naval Observatory, Washington DC, USA.

ANNEX I

AUTHORITIES RESPONSIBLE FOR STATIONS APPEARING IN TABLES I AND II

Station	*Authority*
CHU	National Research Council, Time and Frequency Section, Physics Division (m-36) Ottawa K1A OS1, Ontario, Canada. Attn. Dr C. C. Costain
DCF77	Physikalisch-Technische Bundesanstalt, Laboratorium 1.22 33 Braunschweig, Bundesallee 100, Federal Republic of Germany
FFH	Centre National d'Etudes des Télécommunications, Groupement Etudes spatiales et Transmissions, Département Dispositifs et Ensembles fonctionnels 38, rue du Général Leclerc, 92131-Issy-les-Moulineaux, France
GBR	1. Time information: Royal Greenwich Observatory

Station	*Authority*
	Herstmonceux Castle, Hailsham, East Sussex EN27 1RP, United Kingdom
	2. Standard frequency information: National Physical Laboratory, Electrical Science Division Teddington, Middlesex TW11 0LW, United Kingdom
HBG	Service horaire HBG, Observatoire cantonal CH-2000—Neuchâtel, Switzerland
IAM	Istituto Superiore Poste e Telecomunicazioni Viale Europa, 00100—Roma, Italy
IBF	Istituto Elettrotecnico Nazionale, Galileo Ferraris Corso Massimo d'Azeglio 42, 10125—Torino, Italy
JJY, JG2AR, JG2AS	Frequency Standard Division, The Radio Research Laboratories, Ministry of Posts and Telecommunications Midori-cho, Koganei, Tokyo 184, Japan
LOL	Director, Observatorio Naval Av. Costanera Sur, 2099, Buenos Aires, Argentine Republic
MSF	National Physical Laboratory, Electrical Science Division Teddington, Middlesex, United Kingdom
NBA, NDT, NPG, NPM, NPN, NSS, NWC	Superintendent, U.S. Naval Observatory Washington, D.C. 20390, USA
OMA	1. Time information: Astronomiský ústav ČSAV Budečska 6, 12023 Praha 2, Vinohrady, Czechoslovak S.R.
	2. Standard frequency information: Ústav radiotechniky a elektroniky ČSAV Lumumbova 1, 18088 Praha 8, Kobylisy, Czechoslovak SR

Station	Authority
RAT, RCH, RID, RIM, RKM, RWM	Comité d'Etat des Normes, Conseil des Ministres de l'URSS Moscou, USSR Leninski prosp., 9
VNG	Section Head (Time and Frequency Standards), A.P.O. Research Laboratories 59 Little Collins Street, Melbourne, Victoria 3000, Australia
WWV, WWVH, WWVB	Frequency-Time Broadcast Service Section, Time and Frequency Division, National Bureau of Standards Boulder, Colorado 80302, USA
ZUO	Time Standards Section, Precise Physical Measurements Division, National Physical Research Laboratory P.O. Box 395,0001—Pretoria, South Africa

9

Additional References and Notes

PERIODICALS

Metrologia (Ed. Springer, Berlin and New York) Published under the auspices of the CIPM, contains papers on fundamental problems in the science and technology of measurement, including many valuable results.

I.E.E.E., Proceedings of the . . ., Special Issues on Time and Frequency:

Vol. 54, No. 2 (February 1966);
Vol. 60, No. 5 (May 1972).

I.E.E.E., Transactions on Instrumentation, Special issues on the Conferences on Precision Electromagnetic Measurements (CPEM), No. 4 of even-numbered years since 1960.

CONFERENCES

Conference on Precision Electromagnetic Measurements, CPEM. Held every even-numbered year since 1960. Organized by IEEE Instrumentation and Measurement Group, NBS and URSI. Proceedings see above (IEEE Trans. on I. & M.).
International Conference on Chronometry (CIC) held every 5 years alternately in Germany (1959, 1974), Switzerland (1964), and France (1969), organized by the respective national chronometrical societies.

Annual Frequency Control Symposium, organized by the US Army Electronics Command, Ft. Monmouth, NJ USA Proceedings available from:

No.	Year	Document No.	Obtain from*
10	1956	AD 298322	NTIS
11	1957	AD 298323	NTIS
12	1958	AD 298324	NTIS
13	1959	AD 298325	NTIS
14	1960	AD 246500	NTIS
15	1961	AD 265455	NTIS
16	1962	AD 285086	NTIS
17	1963	AD 423381	NTIS
18	1964	AD 450341	NTIS
19	1965	AD 471229	NTIS
20	1966	AD 800523	NTIS
21	1967	AD 659792	NTIS
22	1968	AD 844911	NTIS
23	1969	AD 746209	NTIS
24	1970		EIA
25	1971	AD 746211	NTIS
26	1972	AD 771043	NTIS
27	1973	AD 771042	NTIS
28	1974		EIA
29	1975		EIA

*NTIS—National Technical Information Service
  Sills Building, 5285 Port Royal Road, Springfield, Virginia 22151
*EIA— Publications Committee, Annual Frequency Control Symposium
  c/o Electronic Industries Association, 2001 Eye Street, NW,
  Washington, DC 20006

Annual Precise Time and Time Interval (PTTI) Planning Meeting, organized jointly by NASA Goddard Space Flight Center, Naval Electronic Systems Command (US Navy) and US Naval Observatory. For information contact: Mr S. C. Wardrip, Code 814.2, NASA Goddard Space Flight Center, Greenbelt, MS 20771, USA.

OFFICIAL REPORTS

BIH—Bureau International de l'Heure Annual Reports. Address: M. le Directeur, Bureau International de l'Heure. 61 Avenue de l'Observatoire, F 75014 Paris, France.

CCIR—International Radio Consultative Committee. Report published every four years. Last issue: XIIIth Plenary Assembly, Volume VII, Standard Frequencies and Time Signals (Study Group 7), Geneva 1974. Published by the International Telecommunication Union, 2 rue de Varembé, CH 1211 Genève, Switzerland.

Index

absorption experiment, 46
  saturated (methane), 59
accuracy
  definition, 44
  evolution of, 3
  values of standards, 61
ageing coefficient, 16, 41
ageing of quartz crystals, 41
  example, 107 ff
aircraft collision avoidance, 11, 217
aircraft flyover, 217
algorithm for time scales, 123
ALGOS method, 124
aliasing, 24
Allan-variance *see* Variance
allocations, of frequencies for radio
  time signals, 182, 221, 227 ff
ambiguity (± 1 count), 142
atomic clocks
  frequency standards, 43ff
  performance, 60 ff
  reliability, 64
atomic frequency standards history, 44
authorities, responsible (list), 250
autocovariance, 18
averages, 16
  of time scales, 122 ff

bias (in error analysis), 44
BIH, 104
BIPM, 103
Breit–Rabi formula, 47

cavity resonator, 53

CCDS (Coordination Committee for
  the Definition of the Second), 103
CCIR (Internationl Radio
  Consultative Committee), 104
  Recommendation **459**, 137
  Recommendation **460–1**, 220
  Report **267–3**, 227
caesium beam resonator, 45 ff
CGPM (Conférence General des
  Poids et Mesures), 103
characteristic table (in logic
  circuits), 78 ff
CIPM (Comité International des
  Poids et Mesures), 103
clocks, 1
  atomic, 43 ff
  concept of, 96 ff
  reading, 106
clock behaviour, abnormal, 124
clock time, 105
clock transport, 179
code, *see* time code
comb generator, 76
  application example, 155
convergence, *see* spectral density
coordinate time, 7
coordinated time scale *see* time scales
coordination, international, 103 ff
counters, measurements, 136 ff
  microwave, 155 ff
counting frequency divider, 77 ff
crystals *see* quartz crystals
crystals, contour shaped, 39
crystal oscillators *see* Quartz crystal
  oscillators
cutoff frequencies, 20 ff